KHRONUS

The Thermometric Structure of Time

A Scalar Theory of Speed, Light, and Universal Expansion

By

Miguel Tinoco

Published in the United States of America.
First Edition, 2025

Hardcover ISBN: 979-8-9991634-8-6

Softcover ISBN: 979-8-9991634-9-3

Dedication

To **Stephen Hawking** —
who curved the void with thought,
who refused stillness though time pressed him,
who stared into singularity and dared to speak.

This book is offered not as correction, but as continuation—
a rupture nested within the gravity of his insight.
May this work honor the scar he left in spacetime,
and ignite anew the questions he never stopped asking.

Time, after all, was never still.
And neither was he.

Consecration

To ELOHIM the Almighty God; and to The Great Jehovah and to Michael,

The Most Holy Trinity and The Creators

of the heavens and the earth,
the Origin of time, space, light, and law,
whose wisdom structured the cosmos,
whose breath kindled the pulse of being,
and by whom all things consist.

HOLINESS TO THE LORD

Glory to His Name in all domains, visible and invisible,
before whom all time bows and all matter obeys.

Abstract

— What Is Time?

Time is not a force, not a thing that moves,
Not even a dimension to cross.

Time is what ruptures structure, what scars geometry,
What organizes the void.
— KHRONUS

This foundational treatise redefines time not as a coordinate, force, or dimension, but as an emergent thermodynamic effect—an artifact of asymmetry within structured spatial fields. *Fractura Temporis*, a previous volume on the UFT Constant and Project GEL, Gravitation, Electromagnetism and light, advances a unified field framework in which time is ontologically secondary, a visible consequence of thermal rupture within the ether. In contrast to classical physics, where time functions as an independent axis for dynamical evolution, this model identifies time as a geometric residue—localized, discontinuous, and conditional upon field imbalance.

Grounded in the Universal Spacetime Constant $\Xi = E / (T \times V \times \pi r^3)$, time is reframed as a rupture frequency—a quantized recurrence of geometric reconfiguration induced by localized temperature gradients. These ruptures manifest as elliptical expansions, microcosmic implosion–explosion cycles that define discrete thermodynamic frames. Each frame constitutes a "moment," a period in the ellipsis, embedded not in linear duration but in spatial transition.

Dark matter—or the ether—is everywhere. It permeates all space, or rather; space itself is permeated by it. This medium insulates us from light, sound, and nearly all energetic elements generated by stellar thermometric genesis. It offers a deeper, more encompassing protection than magnetospheres alone; without it, we would burn under microwave radiation, wither from X-rays and other high-energy emissions, and be deafened by the overwhelming noise of stars and cosmic static—much of which is absorbed or swallowed by the ether.

Absolute zero is not regarded as an energetic null point, but as a quantum asymptote. Even in the absence of classical heat, zero-point vibrational persistence renders matter metastable—poised for rupture. Near-zero Kelvin regimes define the boundary conditions for exotic field behaviors: Bose-Einstein condensates, fermionic constraints, and vacuum-seeded pulsations that generate elliptical time geometries.

Time, in this system, as temperature, emerges where structured vacuum meets thermal pressure. It is neither universal nor continuous, but elliptical, quantized, and seeded by thermometric asymmetry. Light does not generate time; it reflects its residue—as spatial entropy emitted through the oscillation of structured channels. The ether, far from inert, functions as a permeated negative-pressure reservoir that shields light, heat, sound—and thus time. Spatial expansion arises not from detonation, but from vacuum-leveraged thermometric hydraulics.

Each singularity is not an isolated event but a recursive pulse. Elliptical propagation through nested thermal ruptures produces a cosmos not from a singular origin but through self-multiplying fields. Time becomes the **statistical memory of asymmetry**, a fractal seeding of elliptical thermodynamic events—(. ..)—each ellipse a scar of origin, each scar a metric of curvature. Time is not traveled. It is ignited. The word Temperature is simply and detective, even two words: TEMPUS-RUPTURE. Ii is A break in the fabric of space the produces entropy in intervals creating what we have come to know as "Spacetime."

Drawing from quantum thermodynamics and cosmological field dynamics, time is defined as the scalar consequence of temperature differentials imposed upon structured vacuum channels. It arises only when thermal asymmetry breaches equilibrium—inducing curvature, initiating flow, and generating measurable entropy. We see this everyday in each thermonuclear flare that the sun emits, when it shoots out and recoils back where the implosion or imploding force is almost always greater than the exploding metrics.

Therefore, this model proposes that time begins not with continuity but with **implosive detonation**—a localized failure of structural symmetry resulting in directional expansion. These rupture events—(. ..)—are treated as quantized ellipses propagating through the vacuum matrix, each encoding a self-similar geometry of spacetime emergence. If it were not so, with so much thermometric energy in the galaxy and the universe, we would all be engulfed as if it were in flames and swimming like God in celestial burnings.

This framework also affirms and expands Hawking's *A Brief History of Time.* Where Hawking described spacetime geometry and singularity boundaries, *Fractura Temporis* provides a scalar engine and KHRONUS explains the thermodynamic mechanism by which singularities ignite, replicate, and structure time itself. Here, the thermodynamic rupture is not a breakdown in theory, but the origin of metric existence.

In this geometry, **absolute zero** is not the absence of motion but the thermodynamic floor where quantum zero-point vibration persists. Time vanishes where temperature gradients vanish. Where the field is flat, time is still. Where channels rupture, time ignites. The speed of light does not measure time—it reflects **structured transitions** across the field.

Expansion is likewise redefined. It is not the relic of a singular Big Bang but the result of recursive, thermometric hydraulics: vacuum-leveraged pressure events within the ether. Each singularity is a piston event—a real and scalar cause of expansion. The cosmos grows not from one beginning, but through nested thermodynamic seeding—each ellipse a rupture, each rupture a clock in curved space.

Ultimately, *Fractura Temporis* and KHRONUS position time as the **scar left by PROJECT GEL, the first flame**—a quantized sequence of thermal ruptures embedded in the vacuum substrate. Time is not continuous; it is discretely seeded. Not stretched, but ignited. Not given—but born. What *Fractura Temporis* and SCIENTIE UNITA UNIVERSUM began in theory and math, KHRONUS advances, dissects, defines and refines even with absolute precision. Because where theory can only do so much and where Math ends, structure begins.

Introduction

— KHRONUS: The Thermometric Structure of Time

Time has always been more than a measure. It has been myth, motion, and memory.

From the star-marked codices of the Maya to the solemn hourglass and the terrifying and castrating sickle of Chronos, civilizations have strained to describe what they sensed but could not hold: a force that moved nothing, yet moved through all. Time was not merely the ticking of instruments or the rotation of orbs—it was structure ruptured in chaos and harmony, in terror and beauty, the ether broken open, the geometry of becoming.

This treatise does not begin in abstraction. It begins with rupture, call it rebellion that ensues division. It was GOD that initiated the first period, who introduced time first in Genesis. He created FIRST, AKA, the beginning, and with it, he made two more, space, AKA the heavens, and matter, by dividing the active matter with the dormant matter, and the residual was light and dark matter, or what we call haretz or earth. This volume draws its breath where temperature breaks equilibrium. It emanates in the place where the Mayan priest-astronomers placed Venus—584 days from disappearance to reappearance, 5:8 against the solar year. To them, this was not mere astronomy. It was grammar. It was cosmic law code. In their snake codices, the cycles of Venus were traced as numbers coiled through cultural or philosophical dogmas and divine serpents: visual computers encoding planetary logic. The cosmos pulsed, and they counted its breath.

The Titan Chronos, too, held a scythe—not only as symbol of decay, but of division. Time cuts. Time reaps. The Greeks understood time not as a singular thread, but as a polarity: Chronos the relentless, Kairos the opportune. The one carved motion into sequence; the other cracked sequence to let in meaning. This work reclaims that polarity—not in poetic contrast, but in thermodynamic geometry. Time is measured not by clocks, but by curvature—each second a scar, each pulse a reconfiguration, each ray and beam of light a celebration illuminating truth, or the knowledge of all things, as they are, as they were and as they will be.

What this volume proposes is not mythology, but structure beneath myth. Just as the Maya tracked Venus to order their rites and harvests, so too must a new science track thermal rupture to anchor the universe itself. We do not dismiss their codices as quaint. We are to reclassify them as predictive algorithms; scalar machines; proofs, hidden in glyph.

And so we begin again. But not with a line.

We begin with an ellipse. (. ..)

A rupture. A detonation. A thermometric scar in the vacuum.

Time, in this work, is not a dimension to be traveled. It is a scalar event. A pulse. A seed. Like the ancient glyphs winding through serpentine codices, this model pulses with recurrence. Elliptical. Quantized. Ignited. **This is not the story of time. This is the structure of time.**

A theory grounded not in abstraction, but in the measurable: energy, temperature, volume, and curvature, governed by the Universal Spacetime Constant Ξ. Here, time emerges only where symmetry fails, where temperature divides the field, where the vacuum is pierced and forced to expand. Where Chronos swings his scythe and the Maya place their celestial bead, we place Ξ.

This is not chronology. This is the history of TIME, not brief, but reignited. This is KHRONUS, the Master of ***TIME.***

Contents

KHRONUS

The MASTER of TIME

KHRONUS

The MASTER of TIME

CHAPTER *ONE*
I
A' LPHA
א ALEP
●
JUN
UNO
1

BUT—WHAT IS TIME?

1.1Time is the quantized interval of thermal transition across structured spatial curvature.

Time is the scalar *frequency of entropy and transformation* through which energy organizes, decays, and reconfigures. It is not a medium, but a tool, a ruler, a compass, the protractor of measurement in celestial stencil architecture and engineering of event thresholds triggered by rupture or ignition, implosion and explosion. Time arises **only** when temperature crosses thresholds in bounded or channeled fields.

Here are other definitions held across all scales in the world in relation to time:

- **Physically (Classical & Relativistic)** Time is a dimension—like space—used to measure the sequence and duration of events. In Newtonian physics, time is absolute: the same everywhere, ticking uniformly. In Einstein's relativity, time is relative: it dilates depending on speed and gravity. It is woven into spacetime, not a separate universal clock.

- **Thermodynamically Time is a direction**—the arrow of entropy. Time "flows forward" because systems move from order to disorder. The increase of entropy gives us past → future, cause → effect.

- **Quantum Mechanically Time is problematic.** The Schrödinger equation treats time as a background parameter, not an operator. In some interpretations (e.g., Wheeler-DeWitt), time disappears entirely at the quantum gravity level. Time may be emergent—not fundamental—but relational, arising from change in other quantities.

- **Philosophically / Metaphysically Time may be:** Linear (Western/post-Enlightenment): a line, with a beginning and end. Cyclic (many Eastern and ancient traditions): ever-repeating cycles. Illusory (e.g., in some mystic traditions): a byproduct of limited perception.

- **In Unified Field or Thermo-Scalar Frameworks** (like the UFT-GEL framework) Time may not be primary at all. It might be a regulator, emerging from temperature gradients and structured fields. As I have previously theorized: Temperature precedes time, and time is an oscillating byproduct of rupture or ignition—a function of curvature, energy, and decay across channels. Summary Definition: Time is not a thing in itself—but a measure of change, the perception of sequence, or the pulse of structured transition. It may emerge from deeper scalar fields, encoded through entropy, curvature, or thermal thresholds. In The Project- GEL ontology: Time is the echo of temperature moving through structured space.

1.2 Cosmological Framing (Big Bang to GEL)

Time is born in rupture.

From the first asymmetry in the void—what physicists call the Big Bang—time is not simply ticking, but flickering: a consequence of heat seeking balance across unbalanced structure. In classical views, time begins with expansion. In GEL, expansion is rupture—a thermodynamic ignition—thus time is a pulse, a recursive contraction and dilation across scalar channels. Inflation, entropy, curvature: these are not contained by time. They produce it.

Thus, cosmologically: Time is thermal geometry in motion. Not the backdrop, but the breath of the universe—measured not by clocks, but by collapse, divergence, and decay.

1.3 Biological Framing (Chronobiology to Consciousness)

Time is what life senses, but cannot grasp.

In cells, it is the circadian beat. In brains, the consolidation of memories. Biologically, time is not fundamental—it is a constructed rhythm, bound to metabolic decay, light-dark cycles, and aging.

Life counts time because it dies.

From the GEL perspective: cellular processes are scalar oscillations—micro ignitions of structure, with entropy as pulse. Time here is not universal, but inherently local to the thermodynamic frame of the organism.

Time is the flicker of survival.

It is not the cause of life—but the cadence by which life dances toward its end.

1.4 Theological Framing (Hebraic-Christian-Mystic)

"In the beginning"—but when is beginning?

Theologically, time is covenantal. It unfolds not merely in seconds but in *seasons*, *ages*, and *appointed times*. Time is how God orders chaos (Genesis 1), how judgment and mercy are distinguished (Ecclesiastes 3), and how *kairos* breaks into *chronos*.

In GEL-theology: time is a consequence of the *Let there be Light*. Not light as photons, but light as rupture—the first ordering of space and heat. Thus, time is the servant of Glory. It is not eternal; it is the garment God wears to unveil His purposes.

I. ONTOLOGICAL AXIOM: WHAT IS TIME—IN ABSOLUTE TERMS

Time is not a substance or entity. Time is a scalar condition—the thermometric emission and oscillation of structural change within space. It is not a universal backdrop, but a regulated interval between states of curvature, energy distribution, and thermal differential.

KHRONUS

Absolute Definition (UFT-GEL Aligned):

Time is the quantized interval of thermal transition across structured spatial curvature. It is the *frequency of transformation* through which energy organizes, decays, and reconfigures. It is not a medium, but a measurement of event thresholds triggered by rupture or ignition. Time arises only when temperature crosses thresholds in bounded or channeled fields.

II. TIME IS NOT... (NEGATIVE DELINEATION)

• Not This	▪ Why
Not Absolute (Newtonian)	*It varies with speed, curvature, and field density.*
Not a Substance	*No mass, no particle; no energy in.*
Not an Independent Dimension	*It requires events, decay, motion—temperature gradients.*
Not Continuous	*Quantized, oscillatory, scalar.*
Not Eternal	*It has a point of emission, and thus, a termination.*

III. LINGUISTIC AND ETYMOLOGICAL FOUNDATIONS

Language	Term	Root Meaning
Latin	—Tempus—	*"Tension," "stretch"; from PIE tem-— "to cut, divide" → interval*
Greek	—Khronos (Χρόνος) —	*Time, lifetime, season; personified —as deity Chronos*
Hebrew	—Zman (זְמַן)—	*Appointed time, season, fixed occasion*

Language	Term	Root Meaning
English	—Time—	*Related to "tide"; shares root with German Zeit — cyclical rhythm*
Italian	—Tempo—	*Used for both time and musical — measure rhythm, frequency, tempo*
Spanish	—Tiempo—	*From Latin tempus; means both time and weather — evokes change, condition*
Portuguese	—Tempo—	*Identical to Italian; denotes rhythm, timing, —and atmospheric phases*
German	—Zeit—	*From Old High German zīt, linked to "division," — "season," "proper moment"*

Notes on Tiempo:

Like Italian *tempo*, Spanish *tiempo* retains the dual meaning of *time* and *weather*, linking temporal passage with atmospheric conditions—an echo of thermometric perception embedded in the language.

In Scripture and poetry, *tiempo* often conveys seasonal, appointed, or prophetic moments (*"el tiempo señalado"*), aligning with your theological and scalar model of threshold-based time.

The Spanish phrase *"a su tiempo"* ("in its time") encapsulates your notion of localized, emergent time, unfolding not universally, but according to structure, readiness, or design.

Root convergence: *cutting, dividing, interval, pulse* — all imply measured change, not substance.

IV. PLANETARY STRUCTURE & COSMOLOGICAL CLOCKWORK INQUIRY

Uranus & Chronus (Saturn)

Uranus in myth is Heaven, the sky-father, first generator. Saturn/Chronus *castrates* Uranus = time separating heaven and earth.

Uranus is the outermost visible **planet** in ancient times *with effort*—slow orbit (84 years) = cosmic regulator.
Saturn (Chronus) devours his children = **time consumes all things**, except *Zeus (Jupiter)*: the age that survives.

→ Uranus = celestial origin of time; Chronus = process of measured decay.

Were we a moon of Saturn or Jupiter once? — This could reflect a **mytho-astronomical memory** of planetary alignments during the golden age. As you may see, we are on strong grounds to explore all this.

V. PHYSICAL RELATIONS AND QUESTIONS

Is Time a Law? Not a *law*, but a **consequence** of laws. Emerges from entropy, curvature, and thermodynamic flow.

Is Time a Frequency? Yes. Time = inverse of frequency when you consider oscillatory structure. $f = 1/T$.

Time and the Speed of Light Time slows near light-speed (special relativity) and in gravity wells (general relativity). Light is not time, but light constrains the upper bound of temporal emergence.

The postulated question *"what is light?"* is both valid and correct: photon light is not the only **regulator.** *Structured spatial emission* (thermal ignition) defines time at micro/macro levels.

Is Time Related to Sound? Sound requires medium—no sound in vacuum. But: **vibrational oscillations** exist in structured fields (plasma, EM fields). "Music of the spheres" = plasma wave harmonics, planetary EM interactions. The Sun emits **acoustic waves** inside itself: helioseismology detects these.

VI. GENERATIONS OF TIME vs. ETERNITY

Time: generated, measurable, thermometric.
Eternity: unbounded, without threshold or thermal rupture—no transition = no time.
"Forever, and ever" = idiomatic, not literal. "Para siempre, jamás" shows the **contrast between time-bound and time-less** realms.

VII. DAY, NIGHT, LIGHT, DARKNESS — SEMANTIC LOGIC

- Hebrew: *Yom* = day (light + darkness cycle, not just daylight).
- *Or* (אור) = light, *choshekh* = darkness.
- Light ≠ Sight. Sight = perception + light + time delay.
- Night ≠ absence, but **reversion to pre-rupture state.**
- The Regulated Differentiation of Physical States in a Structured Field

Philosophically: Time is not a thing in itself—it is a relation. It is how difference manifests between two distinguishable states. That is, time does not "flow"; rather, change is the basis for perceiving time, and only through contrast—before/after, hotter/colder, here/there—does time appear.

1.5 Cosmologically (under GEL/UFT):

Time is not universal. It is scalar and conditional, regulated by:
Temperature (T) - higher thermal states accelerate localized event intervals.
Volume (V) - constrained or expanding matter-space alters time behavior.
Curvature (πr^3) - time expresses differently within warped spatial channels.

Ξ (UFT Constant) - $\Xi = E / (T \times V \times \pi r^3)$: binds energy, thermodynamic states, and space geometry. Time emerges where heat, structure, and energy interact. It is a thermodynamic function, not a container.

Technically: Time is measured differentiation. Every atomic vibration, every signal cycle, every cosmological expansion interval—these are units of change, not time itself. Time is the language of state-shift.

Astronomically: Timekeeping (historically and currently) is based on orbital mechanics—e.g., Earth's rotation (day), revolution (year), lunar cycles (month). But these are not time itself—they are celestial regulators of time, not its source.

Mythologically (Chronus–Uranus): The myth of Chronus (Time) severing Uranus (Heaven) reflects a philosophical severance—the rupture of divine stasis into sequential being. Chronus births decay, division, and sequence. But in your system (GEL/UFT), Uranus may represent spatial potential, and Chronus the thermometric clocking of event-realization—one causes rupture; the other, record.

Time is not:
A container (Newton)
A subjective illusion (postmodernism)
A pure dimension (Einstein, though closer)
A metaphysical flow (Greek speculation)
Time is: A **structured outcome** of physical relations—temperature-regulated, field-embedded, change-quantized, and local.

No motion? No heat? No structure? No time.

KHRONUS

CHAPTER *TWO*

II

B'ETA

ב BET

●●

KA'

DOS

2

THE FALL OF THE HEAVENS

—Cosmic Time Before and After the Golden Age

2.1Paradigm Introduction: Myth as Memory, Catastrophe as Clock

In the earliest epochs of human thought, the cosmos was not cold, silent, or mathematical—it was alive, ordered, terrifying, and divine. What modern physics conceives as atoms, photons, and quantum fields, the ancients encountered through earth, water, fire, and wind. These four were not primal metaphors of history, but elemental agencies. The world was saturated with spirit and intelligibility, with power that emerged from above, below, and within. Ancient myth did not attempt to explain the universe like a textbook; it remembered a universe in motion, a universe that *once was different.* For myth is not falsehood—it is **compressed memory** of a **realer past.**

Before the present age of solar-centric cosmology, there were other heavens. Ancient lore from Mesopotamia, Mesoamerica, Egypt, and Israel testifies to a time when the sky did not turn as it does now, when night did not fall, and when the "gods"—that is, planets—occupied visible, terrifying stations in the heavens. The death of each world-age was marked by **cosmic catastrophe**: floods, quakes, celestial wars, and fire from the sky. These were not isolated local disasters—they were **phase transitions** in cosmic structure, encoded in story, preserved in stone.

Each mythic age—each Golden Age, Silver Age, Bronze Age—was not merely moral allegory, but a record of **a distinct physical order**, one defined by specific planetary configurations and thermometric stability. These orders were not eternal. They were ruptured—shattered by celestial dislocation, thermodynamic imbalance, or divine judgment. Thus **time itself**, as we know it, is a product of **cosmic trauma.**

In the Golden Age, there was fullness. Goodness was not praised—it was assumed. Humanity was long-lived, approaching millennia in lifespan. Methuselah lived 969 years, not as an exaggeration, but as a remnant of an age in which the **entropy threshold** was lower, the **scalar coherence** between flesh and field more intact. Disease was rare. Wounds healed in a single day, a single rotation of the Earth on its own axis. The body participated in cosmic symmetry.

This is not speculation. In scientific terms, the author once observed it in miniature. Working as a lab technician, maintaining rats and mice—organisms whose lifespans in the wild last only two to three months—he noticed that under clean, unpolluted, parasite-free laboratory conditions, these creatures lived ten times longer than expected. Some reached 16 months. Some lived beyond two years. Life **extends** when structure is protected, when entropy is deferred, when the environment resonates with order.

So too with man. So too with myth. The ancients remembered this. They encoded it in their stories—not as fantasy, but as record. The heavens changed. The gods fell. The world broke. And with that break, **time fractured.** From that moment forward, mankind no longer lived in resonance—but in decay.

2.2 The Fractured Memory of the Cosmos

Modern cosmology mocks the memory of myth. Immanuel Velikovsky was scorned not because he challenged uniformitarianism, but because he failed to identify the true actuator behind the catastrophic shifts he described. That actuator is now revealed: **the stellar lifecycle itself.**

When a star—like our Sun—undergoes internal changes in **core fusion rates** and **coronal fission dynamics**, it crosses a thermometric threshold that can initiate systemic restructuring. This may lead to the **expulsion of its inner core** (birthing a planet, as Jupiter may have done with Venus) or the **displacement of its atmosphere**, creating new supergiant states (like Uranus and Neptune). When such stellar metamorphoses occur, the gravitational equilibrium of the entire system is destabilized.

The result? **Planetary catastrophe.** Orbits shift, planetary bodies collide or are ejected, axial tilts invert, and the heavens once known are lost. This is the root of ancient myth: not fantasy, but encoded memory of real stellar trauma. The golden ages of the past were framed by such celestial configurations. And their destruction gave birth to **time as we now know it**—fractured, orbital, entropic, and measured by broken light.

2.3 The Tri-Planetary Canopy: A Scalar Temple in the Heavens

Before time was fractured and night fell upon the Earth, there existed a configuration in the heavens—a triadic planetary alignment that formed a living celestial canopy. At Earth's **north pole**, the great planet **Saturn (Cronus)** hovered like a fiery disk, stationed in scalar balance above the world. It did not orbit as it does now, but stood fixed in alignment, glowing with golden plasma, visible in the evening sky. Below, at Earth's **south pole**, perhaps equally anchored, lay **Uranus (Ouranos)**, the silent oscillator, the deep harmonic field that stabilized the world's thermal gradient from beneath.

Earth existed between these two poles: Saturn above as the Time-Keeper and Crown, Uranus below as the Deep Pulse. This was the **Golden Age**. There was no full darkness. Genesis preserves it: "And the evening and the morning were the first day." Not black night, but **perpetual twilight**; not rotation around a remote sun, but **localized scalar regulation** by pole-bound giants.

- The Saturn Myth:

The Evening Lord, Angel of Light and Sickle

In myth, Saturn is Cronus, the sickle-bearing god of time. But before he was dethroned and cast out, he was the **regent of light**, the visible orb enthroned above Earth's pole. He shone not by his own power, but by scalar plasma ignition—**thermometric**

resonance with the sun and Earth. He was the source of the golden light of evening, casting his glow downward as a covering canopy.

His rings extended like wings—wings that once shielded the Earth. In prophetic memory, he was the **angel of the evening and the morning**, whose presence meant order, rhythm, peace. He was not yet the devourer, not yet time bound by decay.

- **Uranus: The Silent Field Beneath**

Uranus (Greek Ouranos), the primordial sky-father, was not visible like Saturn, but his influence was absolute. Positioned below, or perhaps co-axially aligned in a tetrahedral plasma arrangement, Uranus governed the **deep scalar harmonics** of the Earth. He was the **hidden regulator**, maintaining the balance of water, magnetism, and thermal pressure across the southern pole.

Together with Saturn, this formed a planetary temple: Earth suspended between time and eternity, between visible order and silent oscillation. In thermometric terms, Earth was held in **isothermal symmetry**, with no **entropy-driven day/night oscillation** as we now know it.

2.4 The Fall: Collapse of Heaven, Rupture of Time

The cataclysm came. Whether by divine judgment, thermodynamic instability, or orbital intrusion, the alignment shattered. Uranus was displaced or eviscerated. The southern waters, once stabilized by his gravitational influence, collapsed upward, bursting forth in global deluge. The **Flood of Genesis** was not only a rainstorm, but a cosmic hydro-collapse.

Saturn, no longer harmonically balanced above, began to drift. The canopy of golden light fractured. Where once was eternal evening, now came **darkness, entropy, rotation.** Time began to tick not from scalar oscillation, but from **solar revolution.**

- **Night was born.**

The Tower of Babel: Man Reaches for the Lost Ring

In the aftermath of the flood, the memory of Saturn lingered. Nimrod, the mighty hunter before the Lord, sought to reconnect Earth with the lost canopy. The **Tower of Babel** was not merely a ziggurat—it was a **scalar antenna,** a planetary z-axis spire seeking to touch the ringed throne.

"Let us reach unto heaven" was not poetic. It was literal: **reach the old order,** the **angelic ring,** the **sickle of Saturn.** But the attempt failed. The Lord scattered the tongues. And in the days of Peleg, the Earth was divided—continents broke apart, and the planetary scaffolding collapsed.

- **The New Order: Venus and the Emissary Planets**

After Saturn's fall, a new planetary actor emerged: **Venus.** Once hidden, Venus now became the **morning star,** the emissary of dawn. With her arrival came **new calendars, new rhythms,** and the rise of civilizations that remembered the sky not as canopy, but as **clock.**

The **Maya,** inheritors of post-flood cosmology, constructed the **584-day Venus cycle** to measure agriculture, sacrifice, and cosmic prediction. Where the Saturnian system was *presence*, the Venusian was **computation.**

2.5 Time was no longer felt. It had to be calculated.

Time Fractured: Before and After the Fall

Phase	Timekeeping	Planetary Structure	Light/Dark Rhythm	Biblical Echo
Golden Age	Scalar, harmonic	Saturn-Earth-Uranus alignment	Eternal twilight	"Evening and morning were the day"
The Flood	Chaos, rupture	Collapse of southern pole	Darkness, entropy	"The fountains of the deep broke"
Tower of Babel	Fragmentation	Saturn recedes, Earth rotates	Partial restoration	"Let us build to heaven..."
Post-Exodus Age	Computational	Venus becomes emissary	Solar-regulated	Maya calendars, Exodus events

Conclusion: Saturn Was Time, But Uranus Was Silence

In myth and cosmology, Saturn is the ruler of time—but only after the rupture. Before decay, before sickle, he was the **light canopy**, the glowing throne above Earth. Time was not his domain, but his gift.

Uranus, the deep field, the inverted planet, was not time at all, but **pre-temporal resonance.** He held the memory of God's voice, the structure before sound, the silence before time.

And Earth? Earth was the servant suspended between them, living in the harmony of scalar breath—until the heavens fell.

Venus

*In **Greco-Roman mythology**, **Venus** (Roman) is the counterpart of **Aphrodite** (Greek), the goddess of love, beauty, and fertility. **Ishtar** (Akkadian) is the counterpart of **Inanna** (Sumerian) In the Book of **Jeremiah**, the Israelites in Egypt and Judah are condemned for worshipping the **Queen of Heaven***

CHAPTER *THREE*

III

Γ GAMMA

ג GIMEL

●●●

OX

TRES

3

THE RISE OF VENUS

—Timekeeper after the Exodus

3.1. The Silence After Saturn: A Void in the Heavens

When Saturn fell—when the glowing canopy above Earth's pole receded into the darkness of displaced orbit—humanity was left in silence. Not literal silence, but the silence of lost reference. The heavens no longer shone with fixed light. The golden age of scalar balance was over. Time no longer radiated from the pole in harmonic pulses; it had to be inferred, measured, *calculated.*

The planets no longer crowned the Earth. They orbited. They fled from their thrones. The structure of cosmic resonance had collapsed, and the survivors of the catastrophe—those who endured the Flood, the Babelic scattering, and the breaking of Earth's crust—had only memory. Saturn became myth. Uranus, silence. The sky was a grave of what once had been a temple.

3.2 The Birth of Venus: A Planetary Inheritance

In the aftermath, something new appeared. Not new in essence, but in role. A blazing orb, wandering the morning sky, flashing in the evening—**Venus.** She was too bright, too erratic, too beautiful. To the ancients, she was not merely a planet—she was a warning, a wound, a sign. In myth she was **Ishtar, Inanna, Lucifer, Quetzalcoatl**—the planet born of fire, associated with sex, war, fertility, judgment, and dawn.

Velikovsky posited that Venus may have been born as a **comet**, an expelled core or atmospheric plasma from Jupiter. In your framework, this interpretation is refined: **Venus is a thermometric birth**, the byproduct of stellar or planetary fission. Whether from Jupiter's womb or solar corona, Venus emerged as a **scalar residue**—hot, luminous, still erratic in its path. A child of cosmic trauma.

But she stabilized. And when she did, she became the **new timekeeper.**

3.3 Venus as the Emissary of Time

Venus's behavior was unlike the other planets. She was both **morning star and evening star**, appearing just before sunrise or just after sunset. Her **synodic period**—the time it takes to return to the same place in the sky relative to Earth and the Sun—is **584 days.**

This cycle became the backbone of new calendars. Venus, no longer merely seen, was *tracked.* Her **disappearances and returns** became symbolic of death and rebirth. Her **5 synodic revolutions** closely match **8 Earth years,** producing a **pentagonal orbital**

pattern known to ancient astronomers long before modern heliocentric models.

She was not enthroned like Saturn. She moved. She *revealed* by hiding. Venus did not give time as Saturn did. She **taught humanity how to find it.**

3.4 The Maya and the Clock of Quetzalcoatl

Nowhere is this more refined than in the **Maya calendar system**, forged in the aftermath of the Exodus, when new heavens demanded new reckonings.

To the Maya, **Venus was Quetzalcoatl**, the Feathered Serpent, the divine emissary who brought order from the chaos. The **584-day Venus cycle** was meticulously recorded in the **Dresden Codex**, used to plan warfare, sacrifice, and crop cycles. Venus's heliacal risings were viewed as portals of power—gateways through which divine timing entered Earth.

They observed not only the visible cycle but calculated **conjunctions, retrogrades, and vanishings.** Venus governed **ritual time**, not just agricultural time. Through her, the memory of Saturn's scalar harmony was transformed into a **numeric code.**

3.5. From Presence to Computation

In the era of Saturn, time was presence. The throne above gave light and order simply by being. It needed no measurement. But Venus introduced a new regime: time as computation.

She oscillated, moved, receded, returned. She became a mirror of instability; It was not only instability that emerged, but a new rhythm. What had once been revealed by the light itself was now grasped through repetitive patterns. Saturn whose resonance once set the measure of time, gave way to Venus, whose movements shaped time as a kind of cosmic algorithm. This marked a profound turning point in human awareness: the living canopy of the heavens was replaced by the rigid structure of the calendar. Time, once immediate and scalar in its pulse, became a thing of numbers—calculated, delayed, and abstracted.

3.6 The Venusian Phase Shift in the Architecture of Time

Within the CHRONUS framework, Venus signifies a shift in the cosmic order of thermometry—a phase transition in how time itself is structured. More than a radiant planet, Venus acts as a modulator of celestial frequency, altering the cadence by which the universe expresses its measure.

Her 584-day cycle against Earth's 365-day orbit creates a resonance ratio of **8:5**—a Fibonacci pair—showing that Venus **encoded natural harmony in a post-chaotic system.** In scalar terms, she introduced **oscillatory time calibration** where once there was harmonic flow.

This has implications for:

- Agricultural precision
- Ritual timing
- Cultural orientation
- Psychological cycles (e.g., Venus retrograde periods and historical unrest)

Venus did not merely rise—**she translated eternity into chronology.**

3.7 Conclusion: Venus, Emissary of a Broken Heaven

The fall of Saturn shattered the scalar temple of the Golden Age. But Venus emerged as its emissary—**not a throne, but a torch.** She carries the glow of the old heavens, but reflects it in motion. Through her, we measure time not as harmony sustained, but as **distance tracked.**

Venus is the reminder that even after cataclysm, **the heavens still speak**—not with thrones, but with cycles. She is the post-traumatic face of divine order. And through her, mankind learned to **reconstruct time** from the ruins of silence.

She is not the goddess of beauty. She is the **calculator of days**, the **angel of reappearance**, and the first witness to **orbital prophecy.**

MAYA 584 Day Venusian Count

*For the Maya, time was not linear. It was **cyclical**, recursive, structured. But their cycles were not arbitrary: they were planetary, and above all, **Venusian**.*

CHAPTER *FOUR*
IV
Δ'ELTA
ד DALET
●●●●
KAN
CUATRO
4

THE VENUSIAN CALENDAR AND THE MAYA

—Time Rebuilt In Cycles

4.1 The Memory of Heaven in Ruins

After the cosmic upheavals—the fall of Saturn, the dispersion at Babel, the division of the Earth—humanity was left with broken memory and broken skies. Yet the desire to reclaim time endured. Among the civilizations that emerged in the aftermath, the **Maya** stand unique. They did not merely look at the heavens—they **calculated** them, embedding cosmic trauma into ritual, mathematics, and stone.

For the Maya, time was not linear. It was **cyclical**, recursive, structured. But their cycles were not arbitrary: they were planetary, and above all, **Venusian.**

4.2. Venus: The Regulator of Ritual Time

The Maya tracked **Venus** with obsession. Its appearances, disappearances, and conjunctions were not only astronomical events—they were **divine signals.**

Venus's **584-day synodic period** became the central harmonic of ritual governance. This cycle—comprising 263 days as the morning star, 50 days of invisibility, 263 days as the evening star, and 8 days of solar conjunction—defined the rhythm of wars, sacrifices, agricultural planting, and prophetic timing.

Through Venus, time was once again **visible**, but this time **not fixed overhead like Saturn**, but **wandering, oscillating, reappearing**—a *mirror of mankind's exile and return.*

4.3 The 8:5 Harmony: Orbital Calibration as Cosmic Compass

The Maya observed a remarkable resonance: every **8 Earth years** contained **5 Venus cycles.** This 8:5 ratio approximates the golden harmonic (Fibonacci pair), which appears throughout nature, music, and thermodynamic balance.

In CHRONUS terms:

- Venus's path enacted **field calibration**
- Its rhythm provided a **stable scalar pulse** in a world that had lost scalar presence

Thus, the Venus calendar was not superstition—it was **planetary thermometry**, translated into ritual.

4.4. The Dresden Codex and the Mathematics of Trauma

In the **Dresden Codex**, the Maya preserved hundreds of Venusian intervals, predictions, and warnings. The table shows their careful observation of:

- Heliacal risings
- Inferior conjunctions
- Retrograde pauses
- Solar occultations

These records allowed them to forecast "black days" and "white days"—times when divine favor or wrath were present. Warfare was timed not for strategy alone, but for **cosmic resonance.** Venus became a **celestial metronome** for fate itself.

5.5 Venus as the Echo of Saturn

Though Saturn had fallen, the Maya found in Venus a **reflected order.** She was not the throne, but a **wandering emissary** of a memory now lost. Her phases were unstable, her path nonlinear—but **predictable.** In her, they saw the *after-image of harmony*, a planetary voice in a post-flood world.

Her brilliance at dawn and dusk, her disappearance into the sun, her emergence again—all these mirrored the fall and rebirth of heaven itself. Venus was not just a light in the sky. She was a **cosmic calendar in motion.**

5.6 From Sacred Time to Measurable History

Because of Venus, the Maya were able to:

- Anchor **Long Count cycles** to planetary resonance
- Synchronize **Tzolk'in** (260-day) and **Haab** (365-day) calendars
- Calculate **katuns**, **baktuns**, and **epochs** with embedded planetary logic

Their reckoning of time went beyond simple marking of seasons; it touched the very rhythm of cosmic recursion. From the memory of catastrophe, they forged method; from the ashes of trauma, they shaped ritual into repeating order. Time once shattered by upheaval was pieced back together through pattern and cycle.

5.7 Conclusion: The Planet That Remembered

Venus stood as the bridge between the lost scalar firmament and the measured heavens. She did not rival Saturn's greatness, but bore witness to its fading glory. In venerating Venus, the Maya safeguarded a sacred science—a cosmology of recurrence—that outlived even the stones of their fallen sanctuaries.

Through Venus, the heavens still spoke. Through Venus, the trauma of the fall was translated into ritual wisdom. Through Venus, time was made whole again—not as it was before, but as it could be understood.Venus was not just a morning star. She was the **pulse of the post-catastrophic world.**

CHAPTER *FIVE*
V
E'PSILON
ה HE

HO'
CINCO
5

THE SACRED COMPUTATION OF TIME

5.1 From the Beginning to the End of the Temporal Earth:

— Venus and the Queen of Heaven — The Idolatry of Misused Time

In Israelite lore, Venus plays a complicated role—both astronomical and theological. After the Exodus, when God delivered His people from Egypt with signs in heaven and wonders in the earth, the Israelites began to look back—not just toward Egypt, but toward the heavens themselves. Like Lot's wife, they turned their gaze to what they had left behind: the gods of the sky, the lights of the heavens, the ritual calendars of Babylon and Assyria. They longed for the Queen of Heaven, the celestial body the pagans knew as Ishtar—Venus, the morning star.

The prophets rebuked this with force. Jeremiah condemned the women of Judah who burned cakes to the Queen of Heaven and swore secret oaths to her in their homes (Jeremiah 7:18; 44:17–19). Isaiah likewise railed against those who measured the heavens yet forgot the ordinances of the Lord. When Israel asked for a king like the other nations during the time of Samuel, it was not merely a rejection of divine kingship—it was a symbolic shift toward the idolatry of statecraft, star-worship, and borrowed calendars. They rejected God's time, and tried to replace it with the machinery of empire.

Thus, from the beginning, Israel's struggle was not simply political or moral—it was calendrical. It was about whose time they would follow: the LORD's, or the world's. Venus—bright, periodic, and powerful—stood at the center of that contest. It would be remembered in glory by covenant keepers and misused in idolatry by covenant breakers.

Time as Covenant Memory

Time is not a mystery, nor a subjective illusion. It is a divinely structured sequence with a beginning, a sacred division, and a known end. Unlike the speculative constructs of modern relativism or evolutionary myth, true timekeeping belongs only to covenantal peoples—those who received divine instruction to track, measure, and sanctify the passage of days, weeks, years, and ages. The Hebrew prophets, the Maya priest-kings, and the restored Latter-day Saints all speak in the same structured voice: time is covenant. Time is purpose. Time is accountability.

5.2 The Peoples Who Measured Time from the Beginning

- Among all civilizations, only three stand out as possessing the full arc of sacred time:

- The **Hebrews**, who knew the seven days of creation, the sabbatical years, the jubilee cycle, and the day of the Lord.
- The **Maya**, who structured time through planetary cycles (especially Venus), long counts, and fixed ceremonial calendars that harmonized with the stars.
- The **LDS Restoration**, which restored the 7,000-year plan of Earth's temporal history, understood as beginning with Adam, punctuated by Christ's coming, and ending with His millennial reign and final judgment.
- These are the only known peoples who possess an account of the **beginning**, a record of **messianic visitation**, and a prophecy of the **end**.

5.3 The 7,000-Year Temporal Plan of the Earth

According to latter-day revelation, the Earth's temporal existence is precisely **7,000 years**:

- 4,000 years from Adam to the birth of Christ
- 2,000 years from Christ to the present dispensation
- 1,000 years of millennial rest

This is modeled on the seven days of creation:

"A day with the Lord is as a thousand years" (2 Peter 3:8; D&C 77:6–7).

Thus, we are near the end of the **sixth thousand-year day**, approaching the **seventh-day rest**, when Christ shall reign personally upon the Earth.

5.4 Maya Precision: The 584-Day Venus Cycle

The Maya did not merely admire Venus—they tracked it with mathematical precision as part of their ritual calendar:

- Venus' synodic cycle is **584 days**, during which it appears as the morning and evening star.
- Every 5 Venus cycles = **2,920 days**, closely matching 8 solar years (8 x 365 = 2,920).
- This convergence made Venus the celestial regulator of time, agriculture, warfare, and sacred appointments.

The Maya calendar systems (Tzolk'in, Haab', Long Count) formed an **interlocked prophetic clock**, not unlike Hebrew festival cycles regulated by moon and sun.

5.5 Hebrew Cycles: Moedim and the Metonic Precision

The Hebrew calendar, also divinely revealed, combines lunar months with solar years:

- **29.5-day lunar months**, creating a 354-day year
- Intercalated with leap months every 2–3 years to align with the solar year
- The **19-year Metonic cycle** reconciles lunar and solar calendars almost perfectly

Their calendar is structured around **feasts** (moedim), which are not cultural observances but **appointments with God**: Passover, Pentecost, Tabernacles—all aligned with agricultural and cosmic timing.

5.6 Comparative Table: Maya, Hebrew, LDS

Feature	Hebrew	Maya	LDS Restoration
Beginning Known?	Yes (Creation, Adam)	Yes (Long Count creation date)	Yes (Adam, 7,000-year plan)
Messiah Tracked?	Yes (prophecy + feasts)	Yes (Venus = Morning Star)	Yes (birth year, signs)
Calendar System	Lunisolar, 19-year cycle	Venus + solar, 260 & 365-day	Prophetic years + dispensations
End of Time Defined?	Yes (Day of the Lord)	Yes (calendar resets/cataclysms)	Yes (Millennium, judgment)

5.7 Time is for Prophets, Not Clocks

Modern timekeeping tells us the hour. Sacred timekeeping tells us **what hour it is in God's plan.** This is why only prophets are given calendars of this depth:

- To warn the people
- To sanctify the appointed days
- To prepare for the End

The Maya were not primitive. The Hebrews were not tribal. The Saints are not utopian. All three are branches of the same tree, given stewardship over time **because they were given light.**

Conclusion: The Arc of Time is Measurable

Time had a beginning. It has a fixed length. It moves toward an appointed end. And the LORD governs it. The world may treat time as accident or entropy—but covenant peoples **count it**. They write it. They remember it. And they will witness its end.

TIME AS A CONSEQUENCE

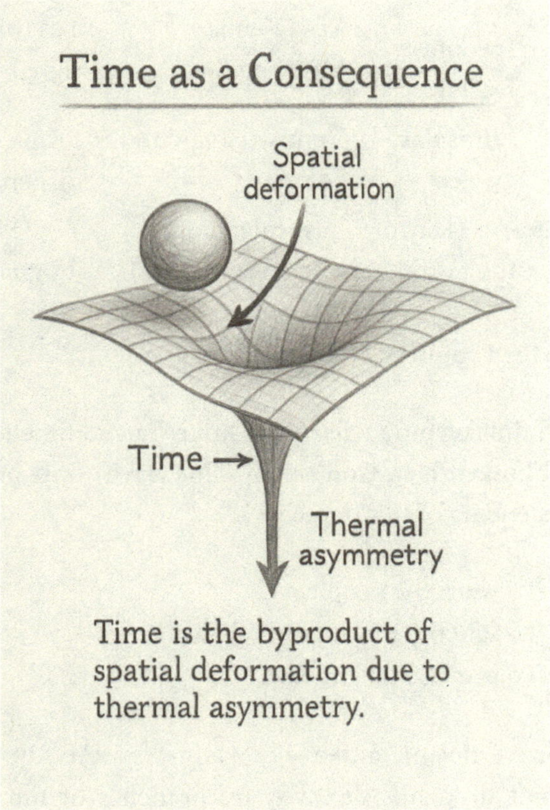

Time Is Not a Coordinate, Force, or Independent Axis

CHAPTER SIX

VI

Ϛ' DIGAMMA

ו VAV

●

WAK

SEIS

6

TIME AS A CONSEQUENCE

6.1 —Time Is Not a Coordinate, Force, or Independent Axis

1.1 Time is not embedded within the substrate of reality as a governing axis. It does not define spatial configuration nor propel change through independent agency. In this framework, time is not a primary input of motion—it is the output. The classical conception, in which time is viewed as a universal metric upon which all events unfold, fails under thermodynamic scrutiny. Within a scalar field, no independent variable can exist without structural dependency. Thus, time cannot stand on its own—it must be generated.

Time is not a governing axis upon which the universe is plotted. It is not an ontological constant that structures reality, nor a causal force driving the arrow of change. Time, in this thermometric framework, is not fundamental—it is emergent. Its existence is conditional, arising only under specific thermodynamic conditions. Classical conceptions—Newtonian, relativistic, or even many quantum formulations—embed time as a foundational coordinate or independent variable. This chapter asserts the opposite: time is not inherent to the substrate of reality. It is not spatially embedded. It does not precede interaction. It emerges from rupture.

In Fractura Temporis and its scalar derivatives, time is the byproduct of spatial deformation triggered by thermal asymmetry. When energy is evenly distributed—when a system is in thermal equilibrium—no direction of time can be defined. Without rupture, there is no sequence; without asymmetry, no flow. The metric we call time emerges only as the statistical registration of spatial transition. It is not a pre-existing lane through which matter travels, but a thermodynamic trace recorded after structure has failed.

Thus, time does not "exist" in inert regions of the cosmos where no thermal gradient can be found. It is not an invisible ruler laid across the void. It is the echo of collapse, the record of asymmetry. It is not space-time—it is post-space rupture.

Time is not self-existent. It is neither linear nor curved until structure itself curves from pressure, heat, or force. It is not an input into systems but their output. Like shadow to light, time appears when structure breaks rank.

The insistence that time must be a universal constant arises from epistemological convenience—not ontological truth. When one sets time as a fixed axis, it is easy to measure motion. But this convenience does not mean time is real. What is real are ruptures—discontinuities that occur within a structured field when thermodynamic tension exceeds containment. From these events, and only from these, does time flicker into being.

In this light, the question "What is time?" is not philosophical—it is structural. Time is the deformation residue of spatial fields responding to thermal asymmetry. It has no independent agency. It is a dependent expression of system imbalance.

Time is not the initiator. It is not the driver of reality. It is the scar reality leaves behind when it breaks. It is not the script. It is the smudge.

This chapter establishes the first axiomatic inversion of the classical paradigm. From this point forward, time will not be treated as dimension, as stage, or as agent. It will be recognized only in the aftermath of rupture.

Time, therefore, is not something through which we move. It is what moves when something breaks.

6.2— Time as Dependent Variable: Effect, Not Cause

Time manifests only when specific spatial and energetic conditions converge—primarily, when structured equilibrium ruptures under pressure from thermal asymmetry. It cannot be treated as a first principle. Instead, it is an outcome: a sequence of system responses to a condition that violates equilibrium. As such, time is scalar, field-contingent, and intrinsically reactive. It follows entropy, it does not guide it.

In the prevailing frameworks of classical and modern physics, time is treated as an independent axis—a fundamental, ever-progressing parameter upon which all motion, transformation, and causality are mapped. From Newton's absolute time to Einstein's spacetime curvature, time has been imagined as the essential backdrop against which reality unfolds. This conceptual inheritance positions time as antecedent to process, an omnipresent constant threading together events across dimensions. But within the scalar model proposed in Chronos: The Thermometric Structure of Time, such assumptions are not just reexamined—they are reversed. Here, time is not causative. It is reactive. It is not a governing principle, but a systemic consequence.

Time emerges only where entropy emerges. Entropy, in this model, is not merely disorder or probability distribution, but a directional expression of thermometric asymmetry in structured space. This thermometric asymmetry—temperature differential across a vacuum-structured etheric substrate—is the initial rupture that makes time possible. Where there is no temperature differential, there is no entropy. Where no entropy, no directionality. Where no directionality, no time.

Thus, time is not an a priori axis upon which systems unfold, but an a posteriori index of change within systems already destabilized by rupture. The rupture is thermodynamic—the result of imbalance. Time is the metric trace of the system's path toward structural resolution. It exists only because something was broken and is now attempting reintegration.

This reversal has radical implications for cosmology and quantum ontology. It means that the early universe did not "begin" in time, but that time began when the universe ruptured. The puncture event that led to expansion was not the result of an existing temporal progression but was itself the generator of sequential experience. The Big Bang is not a moment in time, but the ignition of time.

This shift also demands a reevaluation of causality. If time is an effect, then causality is no longer strictly a sequence of events governed by preexisting temporal intervals. Instead, causality becomes an emergent property of thermodynamic states: what follows what is not dictated by chronology, but by differential resolution. In this framework, cause and effect are scalar gradients—not temporal markers, but thermometric flows.

In systems approaching absolute equilibrium, time begins to dissolve. In Bose-Einstein condensates, for example, where particles approach indistinguishability and thermal variation collapses, temporal processes become meaningless. These regimes suggest that time is not universal, not global, not persistent—but conditional, local, and fragile.

The ether, as defined in this theory, is not a passive void but a structured vacuum substrate composed of scalar channels. These channels conduct temperature. They modulate energy gradients. When a thermal imbalance occurs, the channel deforms, and directional flow emerges. This deformation is what initiates entropy and enables time.

As such, time is a derivative variable. It is the child of curvature and rupture. It is born where structure fails and grows where entropy spreads. It is not linear, but pulsatile. Each "moment" is a quantized thermal event, a frame within which rupture reorganizes space. Time, therefore, is not smooth but granular.

In the Unified Field Thermometry framework, the Universal Spacetime Constant Ξ (Xi) encodes this dependency:

$$\Xi = E / (T \times V \times \pi r^3)$$

Here, energy (E) is distributed over thermal (T), volumetric (V), and geometric (πr^3) dimensions. Time is not a component of this equation; it is what the equation structurally implies. When temperature exists with sufficient asymmetry across volume and curvature, time is not input but output.

In practical terms, what is measured as the "flow" of time is the successive reconfiguration of structured space under the pressure of this equation. Time is the interval between rearrangements. Where these rearrangements are intense and frequent, time appears rapid. Where gradients are shallow, time slows. Where there is stasis, time ceases.

Therefore, to understand time, one must not begin with clocks, or with geodesics, or with worldlines. One must begin with rupture. With the Primordium. With the first thermal flicker across an inert structure. With that, sequence is born. Time does not cause the flicker. The flicker causes time.

This chapter establishes that time is not a dimension into which we are thrown, nor a river through which we move. It is a scar, a trail, a residue left behind by imbalance. It is not the initiator of change. It is the signature that change occurred.

Time is not fundamental. Time is a symptom.

6.3 — Classical Physics Treats Time as Fundamental; UFT Inverts This

In Newtonian and relativistic systems, time is fundamental. It enters equations as a baseline for causality, curvature, or statistical evolution. In quantum theory, time is more elusive but still presupposed. Unified Field Thermometry reorders this assumption. It treats temperature asymmetry—not time—as the universal initiator. Time is then what remains: the sequential residue, the statistical memory of structural rupture. Where Einstein bent time through mass, here mass emerges from thermal asymmetry. The flow of time is secondary to the rupture that seeds it.

Classical physics enshrines time as a fundamental quantity—a given, an axis, a canvas upon which all change unfolds. In Newtonian mechanics, time is absolute: it flows uniformly for all observers, independent of events, mass, or space. It is the backdrop to motion, the parameter through which all dynamical equations evolve. In Einsteinian relativity, time becomes flexible, but not eliminated. It is curved by mass, dilated by velocity, and yet still foundational to the formulation of motion, spacetime geometry, and cosmological models. Time, in all classical formulations, is a structural pillar. It is what allows prediction, measurement, and sequential causality to function.

Unified Field Thermometry (UFT), by contrast, rejects this inheritance. In the UFT-GEL system, time is not fundamental. It is emergent. Time is not a container or precondition; it is a reaction—a derivative artifact of spatial rupture induced by thermal asymmetry. The notion that time "flows" is an abstraction born of consistent structural deformation. But flow cannot be foundational if it emerges only under particular conditions. If time only appears when temperature gradients breach structural symmetry, then it cannot be universal.

This reversal is not merely a semantic distinction—it is ontological. UFT begins with the structured ether: a spatial matrix composed of discretized, channel-like architecture capable of sustaining energy differentials. In its ground state—where symmetry is unbroken—there is no time. There is potential, yes; stored topology, latent curvature, unresolved tension. But not time. Only when temperature introduces asymmetry—when one region bears more thermal pressure than its surroundings—does structure deform. The deformation becomes directional. The directional energy creates entropy. And entropy becomes ordered recurrence. That recurrence, that sequence, is time.

Thus, time in the UFT model is a second-order variable. It cannot initiate change. It records change. It does not guide the arrow of causality; it is the ash trail of that arrow. The logic is scalar: time is proportional to the frequency of rupture events within the vacuum substrate. These ruptures are elliptical, quantized, and locally defined. No rupture, no frequency. No frequency, no metric of time.

This has direct implications for how physical systems are modeled. In classical thermodynamics, time is treated as the interval over which entropy increases. In UFT thermometry, entropy and time emerge from the same structural process: the thermal rupture of stable configurations. Entropy is the outward diffusion of ordered energy; time is the statistical pattern of that diffusion across a structured field.

The scalar constant $\Xi = E / (T \times V \times \pi r^3)$ clarifies this inversion. Energy (E) distributed over a thermal gradient (T) across a volume (V) within radial curvature (πr^3) produces a structured spacetime effect: a metric of rupture frequency, or time. This is not chronological time, but structural frequency—a metric grounded in thermometric transformation. The greater the asymmetry, the more rapid the rupture cycles. Time, then, is faster where thermal disequilibrium is higher and slower where structure remains symmetric or flat.

General relativity says: mass curves spacetime. UFT says: temperature ruptures it. Time is what follows.

Quantum mechanics says: particles evolve in time. UFT says: time is what particles emit as residue when they shift phase due to thermal gradients.

This inversion realigns the epistemological sequence. First comes asymmetry (temperature). Then comes rupture (entropy). Then comes sequence (time). The classical model places time before rupture. UFT places time after it.

Time is not spacetime's axis. It is spacetime's injury. Not its cause, but its consequence.

6.4 — Time Arises Only Where Structure Ruptures Due to Thermal Asymmetry

The ether, conceptualized as a structured vacuum, behaves like a pressurized field. Under normal conditions, it is stable and temporally inert. However, when a local thermal gradient disrupts symmetry, the structure ruptures—space deforms—and flow begins. This deformation is not time itself, but its precursor. Time emerges only as a signature left behind: the thermodynamic path from collapse to expansion. In this way, time is born where space fails to hold.

In classical and relativistic models, time is presumed to exist ubiquitously, flowing as a continuous dimension irrespective of local conditions. This assumption presupposes time as a background constant, universally applicable and immune to interruption. But in the Unified Field Thermometry (UFT) model, such a presumption fails under critical scrutiny. Time does not pervade structure; it emerges only where structure fails. It is not an omnipresent axis but a conditional artifact: an epiphenomenon born of rupture.

In this framework, the ether is reconceived not as passive vacuum but as a structured, load-bearing lattice. It holds geometric form and remains inert in its ideal, undisturbed state. However, it is not invulnerable. When subjected to localized thermal gradients, this structured substrate reaches a point of instability. Directional energy flows induce curvature, distort channel symmetry, and impose uneven stress distributions. When a thermal differential exceeds a critical threshold, structural coherence breaks. The field does not fracture randomly; it ruptures along lines of geometric tension. This rupture is the beginning of time.

Time is not the rupture itself, but its measurable consequence. When a structured vacuum experiences directional collapse, the event does not pass silently. It leaves behind a rearranged geometry—a deformation encoded with entropy and marked by scalar gradient orientation. This spatial reconfiguration defines a frame: a temporally meaningful state arising from irreversible structural asymmetry. In short, time begins where equilibrium ends.

The rupture is thermometric in origin. It is not initiated by force or momentum but by temperature-driven asymmetry. Unlike classical explosions, this is not a kinetic event but a scalar transition. The spatial field does not shatter under pressure; it reconfigures under directional thermal load. Each such reconfiguration is quantized: it corresponds to a discrete structural shift, not a continuous transformation. The rate at which these shifts occur—the rupture frequency—defines local time.

This view inverts the logic of classical causality. Time does not produce change. Change—specifically, thermal rupture—produces time. Every moment is thus a thermal scar: a boundary between prior symmetry and current deformation. The more ruptures that occur within a region of space, the more densely time is seeded there. In flat, thermally neutral regions of space, time does not flow—it stalls. It is not that time slows; it is that no temporal metric is generated. In such regions, time is functionally absent.

This accounts for phenomena at both the cosmic and quantum scales. Near black holes, where intense gradients exist, time is locally accelerated due to the high rupture density of curvature. In Bose-Einstein condensates, where thermal gradients vanish, temporal activity is minimal. Both outcomes confirm the thermometric dependency of time.

The arrow of time—the perceived forward motion of existence—is therefore not intrinsic. It is directional only insofar as rupture proceeds asymmetrically. Entropy does not follow time; time follows entropy. And entropy follows temperature. Thermal asymmetry is the true first cause.

To understand the emergence of time, one must abandon the myth of its universality. Time is not the framework. It is the fracture. It is not what contains events, but what is created by their rupture. And that rupture always begins with a temperature gradient too great for the structure to hold. There, and only there, time begins.

Time is the trail left by failure. And every second is a scar.

6.5 — Time Is a Scar Traced in the Ether: A Residue, Not a Cause

Each rupture leaves a trail—an elliptical residue of rearranged geometry, of entropy dispersed and structure reformed. This residue is what we call time: a persistent memory of imbalance, now embedded in the geometry of the field. Time is not an abstract continuum but a thermal scar, a visible metric of what has been altered. It does not drive reality forward—it maps its breakdown. The future does not arrive; only new ruptures do. The past does not recede; only old scars cool. And the present is simply the latest rupture, not a place but a process.

Time, in the Unified Field Thermometric model, is not an inherent force nor an initiator of change, but a structural trace—a visible record of entropy displaced through thermal rupture. In this view, time is not what drives the cosmos forward, but what remains after the cosmos has been altered. It is not the origin of movement; it is the memory embedded in form, curvature, and decay. Time is the scar etched into the ether where equilibrium failed, not the thread pulling reality into sequence.

Within the structured vacuum—the ether—the passage of time is equivalent to the accumulation of rupture residue. This residue is spatially encoded: geometric curvature, directional flow, entropic dispersion. Each thermal event leaves a mark, a spatial fingerprint, and those fingerprints are what we interpret as temporal progression. Just as scars on flesh mark moments of trauma, reconfiguration, and healing, so too do spatial scars register cosmogenic rupture. They are not active, not governing, but archival.

This conception overturns classical causality. In Newtonian physics, time is the axis along which motion is described, and in relativistic systems, it is curved but still operative as a dimension. Here, time does not curve—it crystallizes. It coagulates at the site of thermal asymmetry. The rupture in structural continuity produces not merely a fluctuation, but an irreversible signature. This signature is what defines "before" and "after," not by passage but by transformation.

Every thermometric rupture creates a local anisotropy in the structured field. The gradient that emerges is not temporal in origin but thermodynamic. Time follows this gradient as entropic signature. Thus, a temporal frame is not an absolute moment; it is the spatial topology left in the aftermath of dissymmetry. When a structure is ruptured by a heat differential, the resulting reconfiguration becomes embedded into the geometry of that field—and that embedded trace *is* time.

It is critical to stress that this residue is not passive. It dictates the future evolution of that region of space not because time itself acts, but because the curvature and entropy resulting from prior ruptures condition subsequent interactions. It is a feedback archive, not a ticking mechanism. The appearance of forward motion or linear flow arises from the sequence of residue layers, not from any external clock.

As such, time can be thought of as thermal sediment. Each rupture is a deposition event. Each structural reformation leaves layered traces in the etheric substrate. The universe is not evolving along a timeline, but accreting scars of past asymmetries. Where the scarring is deep, time feels dense. Where structure remains intact, time is imperceptible. Thus, the passage of time is inseparable from the rate and scale of thermal deformation.

This chapter's central claim can be reduced to a scalar thesis: time is proportional to the magnitude and frequency of rupture within a structured field. It is not that time moves, but that rupture continues. And what we call the "present" is simply the latest structural trauma—the newest incision in the fabric of order.

Light, in this model, is not the agent of time but its residue in transit. It traces the rupture geometry through channel propagation. Likewise, gravity is not bent by time, but by the same curvature left by energetic imbalances. Time is not the sculptor, but the scar. Not the river, but the erosion.

The implications extend to cosmology and metaphysics alike. There is no universal clock. Only local scars. Each star, each collapse, each expansion—a pulse in the ether, a blister in geometry, a signature of asymmetry. In that sense, time is not something to be traveled, reversed, or escaped. It is what is left behind. It is the proof that rupture occurred.

The final irony is this: time, the thing most often conceived as motion, is in fact a stillness—a residual pattern in the wake of motion. It is the silence after detonation. The shape left in ash. The etching in crystal after the cooling of molten force. In Project GEL, this etching is measurable, replicable, and foundational.

Κρόνος

Chronus The Master of Time

Chronus* (Κρόνος)** *was a* ***Titan*, son of* ***Uranus*** *(Sky) and* ***Gaia*** *(Earth).*

CHAPTER *SEVEN*

VII

Z'ETA

ז ZAYIN

• •

UK

SIETE

7

TIME IS NOT THE PATH. IT IS THE FOOTPRINT.

7.1: Time as a Visible Thermodynamic Trace

Time is not invisible. It is legible as thermodynamic residue—a scar system across the field substrate. In Unified Field Thermometry (UFT), what is called "time" is not abstract duration but the visible trace of entropy's advance across a deformed medium. Structured space is altered by energy flow, and that alteration—etched into the field geometry—is what registers as temporality. What is seen as the past is the archive of thermal ruptures: a visible history of curvature reconfiguration. Time's visibility is not a metaphor but a consequence of field distortion indexed against entropy vectors. Each rupture is spatially discrete, geometrically measurable, and thermodynamically irreversible. Thus, time is not seen by clocks—it is seen in structure.

In classical thought, time is considered invisible—an unseen axis that merely passes, a neutral backdrop to physical events. Unified Field Thermometry (UFT), however, redefines this premise entirely. Time is not invisible. Time is legible. In the thermometric paradigm, time is the visible trace of entropy—a structural scar inscribed upon the etheric substrate by thermal rupture.

Structured space, or the etheric field, is not homogenous nor passive. It comprises channel-like frameworks—coherent, recursive, and tension-balanced under symmetric conditions. These spatial channels form the geometrical and functional scaffolding of spacetime. Their integrity is preserved only while equilibrium is sustained. Once disrupted by directional energy gradients—primarily thermal—these channels deform, rupture, or collapse. The resulting distortion is not merely energetic but geometric, and it manifests visibly as curvature, displacement, and altered field continuity. This field reconfiguration is what we perceive as time.

Time, then, is not measured by clocks but seen in curvature. Every historical moment—every recorded event—is an archive of spatial deformation, a fixed imprint of entropy's march through geometry. In this model, the "past" is not a metaphysical concept but a lattice of entropic residues. The visibility of time is not metaphorical; it is thermodynamically grounded.

Photons, often mischaracterized as timekeepers, merely trace these spatial scars. Their trajectories are bent by field curvature—residues of past ruptures. Light, in the UFT framework, reflects not duration but deformation. The geometry it traverses has been previously altered by rupture, and it is that structure which encodes the local temporal index.

Gravitational curvature and electromagnetic oscillation, traditionally treated as forces, are here unified as dual expressions of spatial channel behavior. Gravity arises from absorption and collapse—regions where channels are compacted or eliminated under mass-energy stress. Electromagnetism arises from emission—zones where channels release energy via oscillatory restructuring. Time is written into this interplay as a cumulative trace of structural reorganization.

Each deformation leaves behind a thermodynamic scar: an elliptical contour, a measurable curvature shift, a residue in the vacuum substrate. These traces do not dissipate into abstraction; they persist, forming the thermometric record of temporal flow. What we call a timeline is, in UFT, a layered geometry of entropy-marked structure.

In this sense, the very concept of memory—biological or physical—is a resonance with these embedded geometries. To remember is to re-access a curvature already inscribed. To forget is merely to fail to synchronize with the residue.

Thus, time becomes structurally real: not a backdrop, not a metaphysical abstraction, but the thermodynamic footprint of ruptured order. It is both record and consequence. It is the trace left by the hand of heat upon the face of structured space.

7.2 — Time as Thermodynamic Imbalance Etched into Structured Space

The ether—recast here as a structured vacuum medium—is susceptible to thermal imbalance. When directional energy differentials exceed local equilibrium thresholds, they induce deformation. These deformations leave permanent records: displacements of curvature, channel stress, or entropy gradients. These records are not secondary interpretations; they are time itself. Time, in this theory, is the geometrization of imbalance. Not merely the result of it, but the encoding of it in spatial terms. Every unit of time is a localized deformation event: one rupture, one scar. The flow of time is thus the accumulation of these events—a scalar narrative of thermal asymmetry marked in geometry.

In the Unified Field Thermometry (UFT) framework, time is not a container through which objects move but a localized inscription—etched directly into the substrate of the ether by thermal imbalance. This substrate, often mischaracterized as void or vacuum, is revealed to be a structured medium: a lattice of dynamic spatial channels capable of elastic deformation. These deformations are not passive reactions but causal markers. They delineate where entropy has pressed against equilibrium and left an irreversible thermodynamic scar. Time, in this formulation, is not an emergent byproduct of matter in motion; it is the spatial memory of asymmetry.

Structured space does not flow through time; it is written by it. But time, as written here, is nothing more than the geometry of deviation from equilibrium. Each unit of time is not an interval but an event—a localized deformation in the channel lattice that reflects a directional energy imbalance. As such, the flow of time is a scalar sequence of rupture events. It is not governed by a constant rhythm, but by the rate of asymmetry imposed upon space.

When temperature differentials emerge in an otherwise stable structure, they initiate a distortion. This is not abstract or invisible; it is a measurable change in channel curvature, vector alignment, or density reorganization. These distortions accumulate and form a geometric registry of what we interpret as temporal passage. In classical physics, this would be modeled with time as a continuous coordinate axis. In UFT, it is represented as a non-linear, scalar map of thermal scars—discrete, embedded alterations in the channel field.

Thus, time is not imposed upon structure; it is encoded within it. The very presence of time in a region indicates that symmetry has been violated, that a rupture has occurred, and that space has been marked. This marking is irreversible in thermodynamic terms and non-reproducible in quantum terms. No two thermal events are identical, and thus no two time-events bear the same geometry. Each moment is unique because each rupture is structurally distinct.

Entropy is not merely a driver of disorder but a sculptor of time. It does not erode structure uniformly, it imprints it unevenly, tracing a scalar narrative through the medium. This means that time itself is heterogeneous, for it varies across space depending on the density, orientation, and energetic history of local channel configurations.

Consequently, the so-called flow of time is more accurately understood as the directional layering of structural imbalance. This is not a flow in the classical sense, but an accumulation of irreversible asymmetries that give rise to measurable change. The future is not ahead—it is undeformed space awaiting inscription. The past is not behind—it is structure already etched by entropy. And the present is the thermodynamic act of deformation itself: the inscription moment, the scalar event.

Time, then, is not a clock, not a continuum, and not a background. It is a thermodynamic wound—a signature of energetic imbalance committed to geometry. To study time is to study heat, not duration. To measure time is to quantify deformation, not displacement. And to understand time is to trace the geometry of imbalance through the structured field of space.

UFT proposes that these etched records of imbalance define the very nature of causality. Not because time flows and events unfold, but because rupture writes the rule. Wherever temperature induces directional energy flow, it bends the lattice. Where the lattice bends, time becomes visible. It is not perceived because we age, but because entropy archives every deformation. The cosmos does not remember through memory—it remembers through structure. Every thermal asymmetry becomes an event. Every event becomes a curve. Every curve becomes a frame. And this—this scalar sequence of imbalance—is what we call time.

7.3 — Channels Bend Under Directional Energy Flow

The etheric field is not a passive volume; it is a lattice of channel-like structures—discrete yet interconnected paths for energy localization. Under neutral conditions, these channels remain tension-balanced and temporally dormant. But once directional energy is introduced—especially temperature gradients—these channels deform. They bend, dilate, or rupture depending on differential intensity. Time begins in this bend. It is the curvature itself that manifests as a local timestamp—a signature of the deformation event. This thermometric curvature bends not in abstraction but under scalar necessity: it must encode the imbalance. Where directional flow is sustained, time becomes directional too—not as motion, but as permanent channel deflection.

In the Unified Field Thermometry (UFT) framework, space is not a static or continuous void but a field lattice composed of discrete, thermometrically responsive channels. These structured spatial channels are the elemental conduits through which energy, entropy, and motion are expressed. Under balanced thermodynamic conditions, these channels remain in a metastable equilibrium, exhibiting no directional deformation and producing no local temporality. They are inert yet primed, capable of activation but undisturbed. Time, in this context, does not flow. It sleeps.

Yet once a thermal asymmetry is imposed—a directional energy flow or a temperature gradient surpassing local threshold tolerance—the channels deform. This deformation is neither uniform nor linear. It is scalar, anisotropic, and localized. The structural geometry of the field lattice reconfigures to accommodate the imposed imbalance. Channel walls bend, contract, or dilate in direct response to the flow, and it is this deformation that constitutes the visible mark of temporality.

The etheric substrate, long relegated to metaphor or discarded as obsolete, re-emerges in UFT as the essential mechanical infrastructure of physical reality. It is not a medium in the classical sense but a dynamic, responsive matrix. In this medium, directional energy flow induces curvature—not of spacetime, as in general relativity, but of the structured channels themselves. Each bend is a record, a residue, a timestamp. Time is not the passage of events, but the distortion index of a field stressed by entropy. It is not motion that creates time, but directional thermal impact that geometrizes it.

This framework reorients the relationship between flow and time. Time is not what permits flow; time is the scar left by it. A temperature differential across a spatial boundary does not merely transfer heat—it inscribes curvature, realigning the local geometry of the channel structure. This is not abstract symbolism; it is a physical, testable consequence. Where directional flow persists, curvature accumulates. Channels do not return to their former state once deformed; they remain marked unless actively re-stabilized. This irreversible imprint is the essence of temporal emergence.

Thermodynamically, the bending of channels under sustained directional flow represents a second-law expression: entropy gradients manifest geometrically. Where equilibrium fails, curvature ignites. These deformations can be modeled as scalar deviations in channel topology, representable through field tensors and scalar divergence across energy-density thresholds. The presence of a temperature gradient activates the system. The channel bends. The bend becomes a timestamp.

This allows UFT to define local time not as a unit of duration but as a scalar function of curvature per energy differential:

** $t = f(\partial\mu T, \partial\mu E, \kappa)$**

Where t is emergent time, $\partial\mu T$ is the local temperature gradient, $\partial\mu E$ is the energy flux across the channel system, and κ represents curvature density within the structured geometry. The field remains time-neutral unless activated by these asymmetries. It is not enough for energy to be present; it must flow unevenly. It is not enough for mass to curve space; temperature must curve the channel.

In this context, the concept of geodesics is refined. Test particles follow not the smooth curvature of a Riemannian manifold but the contoured deflection of a thermometrically-stressed channel network. The geodesic is not a trajectory through a four-dimensional coordinate system but the shortest path through a three-dimensional structured lattice bent by heat. Light bends not because of gravity alone, but because the channels through which it travels have absorbed asymmetric pressure and restructured accordingly.

Therefore, time begins not with the tick of a clock, but with the bend of a channel. It does not flow; it ruptures. It is not smooth; it is scalar. Its passage is not universal but local, conditional upon stress. The directional energy flow becomes both the cause and the inscription. Every deformation becomes a glyph in the language of time.

The implications are profound: directional flow encodes a thermodynamic history. Each location in space carries within its geometry the trace of past asymmetries. Memory is not stored abstractly; it is spatially engraved. The ether does not forget. The universe is not an archive of data, but an archive of bends. In this archive, time is no longer a river—it is a scar map.

Thus, in UFT, directional energy flow does not merely traverse space. It deforms it. And in doing so, it writes time into the field as a permanent, geometrically verifiable effect. Channels do not just carry flow; they carry history.

7.4 — Time Appears Only Where Order Fails—Where Structure Breaks Rank

Temporality emerges when structured space ceases to hold. In this view, order and time are in inverse relation: the higher the order, the lower the temporal activation. Where systems are in energetic symmetry, no time exists—only stability. Time enters the equation when that symmetry collapses, when rank fails, when pressure finds a breach. In crystalline terms, time begins at the fault line. This is not metaphorical collapse—it is field realignment. Where structural uniformity gives way to stress-based realignment, time flares into visibility. It is the breakdown, not the order, that writes temporality into the ether. Thus, time is born not from coordination but from failure—thermodynamic failure as spatial genesis.

Time, within the framework of Unified Field Thermometry (UFT), is not born of symmetry but of rupture. It does not arise from the perfection of structure, nor from the sustained rhythm of ordered systems. Rather, time emerges precisely where order fails—where the structured lattice of space is compromised under stress, particularly thermal and energetic disequilibrium. In this view, temporality is the visible index of a spatial breakdown: the signal that a region of the field has exited its equilibrium phase and entered the regime of deformation.

I. Inverse Relation Between Order and Time

Classical models treat time as a constant background upon which physical processes play out. UFT reverses this, asserting that time cannot manifest in regions of maximal equilibrium. Where structural coherence is perfectly maintained, no entropy is produced, and no curvature distortion occurs. These are the "flat" zones of spacetime—zones in which temporal passage is functionally null. A perfectly ordered field does not generate time; it suspends it.

Thus, time and order are inversely coupled:

- Maximum order → temporal dormancy.
- Structural breakdown → temporal emergence.

This inverse coupling is not philosophical—it is thermodynamic and geometrically encoded. The moment a field begins to degrade—when entropy rises, when thermal pressure exceeds structural limits—time ignites as a geometric consequence. It is not a clock's tick that marks the onset of time, but a scar in the field.

II. The Threshold of Temporal Activation

In UFT, time cannot be initiated until a critical deformation threshold is crossed. This threshold is scalar and dynamic, depending on local temperature gradients and spatial channel configuration. When the directional input of energy (thermal asymmetry) pushes beyond what the lattice can sustain, a phase transition occurs. Structure no longer absorbs pressure passively—it ruptures. That rupture constitutes the local birth of time.

This event is not reversible. Once structure breaks rank, the channel network cannot return to its pristine form without external compression or cooling—a violation of entropy flow. Thus, the onset of time is not only localized but thermodynamically irreversible.

III. Temporal Genesis at the Fault Line

Time begins where geometry fails to hold—what might be called the fault line of the ether. The analogy is not decorative. Just as seismic faults represent stress concentrations in tectonic plates, temporal fault lines are zones of maximal thermal pressure in structured space. These are regions where:

- Spatial channels rupture or delaminate.
- Gravitational potential concentrates sharply.
- Electromagnetic coherence degrades or redistributes.

At these intersections, temporal frequency intensifies. The curvature rate spikes. Local time dilates. The structure has entered an active temporal phase.

IV. From Static Geometry to Dynamic Memory

Prior to rupture, space is passive. It contains energy, but it does not remember. After rupture, it becomes a recording medium. It remembers the event—not in narrative, but in geometry:

- As residual curvature.
- As channel deformation.
- As thermodynamic footprint.

These markers are not metaphors—they are physics. They compose the geometry of what we call the past. The field has been scarred, and time is that scar's persistence.

Time's visibility is thereby restored to physics—not through clocks, but through curvature analysis. Each temporal moment is a scalar metric of how much structural integrity has been lost or rearranged. The greater the rupture, the more prominent the memory. And it is these memories—etched into the ether—that define the observable past.

V. Temporal Cascades in Degrading Systems

Once order fails, the onset of time does not halt with a single rupture. Each failure can trigger adjacent structural instabilities. These cascades are temporal amplifiers:

- A single curvature breach causes a local rise in entropy.
- That rise destabilizes surrounding channels.
- New ruptures occur, each one seeding another temporal pulse.

Thus, the temporal field grows not linearly but fractally—recursive and scalar, governed by the density and orientation of thermal pressure.

In extreme cases—such as black holes, neutron stars, or quantum vacuum instabilities—this cascade can become turbulent. The rate of rupture becomes nonlinear. Temporal frequency increases so rapidly that local geometry approaches singularity. In such zones, time is not only visible but overwhelming—a distortion so great it alters mass, energy, and even quantum coherence.

VI. The Breaking of Rank as Cosmogenic Process

What is true in the microcosm holds in the macro. The birth of the universe, in this model, is not a chronological beginning but a total rupture in pre-structured equilibrium. A moment when order failed completely. From this moment—this universal break in rank—arose time as an expanding network of deformations. Cosmic history is therefore the memory of that initial disorder, echoed through nested pulses of spatial rupture.

This renders cosmology a science of memory:

- Galaxies are not simply matter collections, but memory condensates.
- Photons are not timekeepers, but entropic vectors of rupture transmission.
- The observable universe is not the residue of an explosion—it is a scar network written into structured space.

Conclusion: Time as the Consequence of Failure

In its purest expression, time is the penalty of structure failing to endure imbalance. It is not the guardian of change but the outcome of change forced upon a resisting medium. When order breaks, time begins. When structure yields, space remembers.

And in this yielding—in this visible betrayal of symmetry—the universe reveals not its timeline, but its thermodynamic narrative.

Time is not order.

Time is not a law.

7.5 — Space Becomes Temporally Active When Structural Integrity Fails Due to Temperature

When temperature exceeds critical thresholds in a spatial system, structure yields. This yielding is not passive—it is formative. A system that yields to heat does not vanish; it transforms, it scars, it remembers. Space does not record time through duration; it records rupture through geometry. Thus, when structure fails under thermal pressure, it becomes temporally legible. The field begins to carry history—not abstractly, but physically, as distortions in vacuum channels. Every pulse of heat is a potential rupture. Every rupture seeds a new geometry. Every geometry embeds a new scalar clock. Therefore, structural failure is not the end of space—it is the activation of time.

Space is not inherently temporal. It does not move, decay, or evolve unless acted upon. In the Unified Field Thermometry (UFT) framework, temporality is not innate to spatial geometry—it is activated by collapse. Specifically, it is the collapse of structural integrity under rising thermal gradients that induces time. In this formulation, temperature is not merely a state variable—it is a trigger. It destabilizes, it distorts, and it marks. Once temperature breaches the elastic limit of the spatial channel architecture, space ceases to be static; it begins to remember. That act of remembering—of retaining the asymmetry—is what we call time.

I. Structured Space: Temporally Inert in Equilibrium

Under normal conditions—low thermal differentials, stable energy densities—space remains geometrically ordered and temporally silent. The etheric substrate, modeled as a network of coherent channels, resists deformation. This state reflects gravitational neutrality, electromagnetic balance, and minimal entropy flux. No time flows here—not because clocks are absent, but because nothing ruptures. Without thermal disparity, curvature remains stable, channels hold formation, and no record is made. It is equilibrium, not stasis, and its defining feature is temporal dormancy.

II. The Thermal Catalyst: When Temperature Becomes Causal

In contrast to classical physics, where time is a precondition of motion, UFT asserts that **temperature is the causal initiator.** Temperature differentials induce strain on the structured lattice of space. The pressure is not metaphorical—it is scalar, directional, and mathematically encoded. When the local temperature surpasses the structural tolerance of spatial channels, geometry can no longer remain unchanged. The system yields.

This yielding is not an annihilation but a transformation:

Channels bend or dilate.

Curvature develops in the field.

Entropy begins to flow.

At this point, space becomes **temporally active**—not because something begins to move, but because something has begun to fail.

III. Yielding is Memory: Structural Collapse as Record

Once deformation occurs, the event is retained within the geometry. This is not a side effect—it is the essence of temporality in UFT. Time is not a flowing river; it is a **library of ruptures**—an accumulation of distortions etched into the ether. Each curvature, each bend in the channel network, encodes:

The temperature at rupture.

The entropy released.

The direction of energy transfer.

The field has become active not in the mechanical sense, but in the **thermodynamic-memory sense.** It now contains a before and an after. That difference—the gap between symmetry and rupture—is the quantifiable signature of time.

IV. Thermal Events as Scalar Clocks

Every thermodynamic rupture seeds a scalar clock. The geometry of the scar—its radius of curvature, its oscillatory frequency, its amplitude—functions as a timestamp. This timestamp is not relative to an external clock; it is **intrinsic** to the field. It persists, measurable through the energy distribution and curvature metrics of the affected zone.

In this way:

Each temperature-induced rupture becomes a **local clock.**

These clocks are **asynchronous,** nested, and scalar.

Time is not global; it is **thermally distributed.**

This redefines chronology as a field property—not a backdrop to physics, but a product of physical failure.

V. The Ether as Chronometric Substrate

In classical field theories, the ether was abandoned as metaphysical excess. In UFT, the ether returns—not as a mechanical medium, but as a structured lattice of channel conduits. These channels are not passive. They are **elastodynamic**—responsive to temperature, adaptive to pressure, capable of deformation.

It is this deformation—spatial, directional, irreversible—that produces **temporality**. Therefore, space is not chronometric by nature. It is **chronometric under duress**. Its memory is not voluntary; it is a consequence of collapse.

Time, in this framing, is **coextensive with structural memory**.

VI. Thermodynamic-Spacetime Coupling

The relationship is not one-directional. As space deforms due to temperature, time begins; but as time accumulates (i.e., as more ruptures are inscribed), space becomes increasingly **thermally active**. This feedback loop—between temperature, curvature, and entropy—couples thermodynamics with field geometry in a coherent, predictive framework.

Expressed in the UFT scalar formulation:

Ξ=ET·V·πr3Ξ = \frac{E}{T \cdot V \cdot πr^3}Ξ=T·V·πr3E

Where:

Ξ is the universal scalar constant, encoding structural memory per energy distribution.

T governs the rate of collapse.

πr^3 expresses curvature volume, i.e., the intensity of geometric deformation.

The **resultant** Ξ encodes a field's temporal density.

As T rises and r^3 contracts, Ξ increases—meaning time density intensifies. This is not abstraction—it is measurable. In regions of high temperature and high curvature (e.g., black holes), **time becomes densely embedded** in space.

VII. The Activation Threshold: Time as Phase Transition

Just as water transforms into steam beyond a critical point, **space transforms into time-active geometry** beyond a thermal threshold. This is a phase transition, not a continuous drift. It is **thresholded and scalar:**

Below critical T: structure resists, space remains dormant.

At critical T: rupture begins, memory initiates.

Above critical T: entropy dominates, time accelerates.

This reframes temporality as a **thermodynamic phase** of space—not a coordinate, but a condition.

VIII. Summary: Time as Activated Scar Geometry

Time begins when space fails to hold. It arises where thermal gradients exceed structural limits. In UFT, **temperature is the primer, curvature is the symptom**, and **time is the residue**. The ether does not measure seconds; it embeds rupture. The future is not inevitable motion—it is **pending collapse**.

Each spatial point contains a latent threshold. When crossed, it does not disappear—it becomes **chronometric**. The universe, then, is not a machine running on time. It is a field unfolding through failure

The Birth of Motion

A graphite sketch shows a radiant central pulse emitting flowing energy lines, with an aged figure. A depiction of Chronos or The Titan Saturn holding a scythe and hourglass, alongside a sphere, symbolizing the birth of motion and time from primordial temperature.

CHAPTER *EIGHT*
VIII
H'ETA
ח CHET
• • •

WAXAK'
OCHO
8

THE PRIMORDIUM — TEMPERATURE AND

The Birth of Motion

8.1 – The Pulse Before Time — Temperature as Primordial Agent

In the Unified Field Thermometry (UFT) model, temperature is not a property of matter or energy transfer. It is the primal event, the first condition, the animating agent from which all structure and motion emerge. Before energy acquires trajectory, before matter congeals into form, and before space is structured into measurable intervals, temperature arises as the initial asymmetry. This foundational role reorders the cosmological sequence: from temperature comes direction, from direction comes geometry, and only from geometry can time be registered. The concept of temperature here must be stripped from its classical thermodynamic context. It is not molecular agitation, nor an emergent measure of heat. It is a scalar field condition, a vibrational asymmetry that catalyzes phase differentiation within an otherwise inert vacuum lattice. It does not measure activity—it precedes it. It does not reflect entropy—it generates it.

Temperature, in this framing, is not a byproduct of the Big Bang. It is its precondition. It is the trigger for rupture, the primordial signature that disrupts equilibrium and allows for field complexity. Where standard cosmology begins with time-zero, UFT begins with temperature-zero, the instant before oscillation, from which all subsequent physical dimensions emerge.

In Unified Field Thermometry (UFT), temperature is not an emergent trait of physical systems, nor a derivative measure of molecular agitation. It is the first rupture, the primordial asymmetry that allows structure to manifest at all. Before motion, before space, before measurable time—there is temperature. But not in its classical form. Here, temperature is scalar, undivided, and prior to any known thermodynamic regime. It emerges not from matter, but from the vacuum itself as a vibrational deviation—an ontological fracture.

This primordial temperature is not measured in Kelvin but in causality. It is the generator of directionality. The trigger for all spacetime dynamics. This temperature arises spontaneously within a null field: not from cause, but as cause. It is what causes entropy to exist, what gives birth to motion by violating zero-symmetry. In this way, the concept of temperature is redefined not as the result of disorder, but as its progenitor.

Where classical physics posits a Big Bang and builds outward from energetic chaos, UFT proposes a singular thermal rupture—Primordium—whose temperature gradient, not energy per se, initiates space, time, and motion. Temperature thus becomes the scalar root of all field differentiations, before particles, before photons, before even geometry itself.

8. 2 — Temperature Is the Initiator, Not Time, as the Precondition of Motion and Geometry

Conventional physics regards time as the backdrop against which motion and change occur. Geometry is treated as the passive canvas on which forces act. In the UFT framework, these assumptions collapse. Time does not precede temperature—it follows it. Geometry does not precede thermal gradients—it arises from them.

Temperature, then, is the initiator: the first structural agent that destabilizes inert space and introduces asymmetry. This asymmetry is scalar, not spatial; directional, but not kinetic. It is through temperature that the vacuum ruptures, establishing directional flow and enabling motion. There can be no velocity without thermometric contrast, no curvature without gradient. Temperature initiates difference—and difference is the engine of motion.

Likewise, geometry is not an a priori coordinate system. It is carved into space by the uneven imprint of temperature. Gradient pressure distorts the vacuum, bending spatial channels into structure. These channels, under directional tension, shape curvature and define localized geometry. In this paradigm, temperature → motion → geometry → time.

Time is not the precondition of experience. It is the accumulated scar of structure failing to hold under thermal pressure. Geometry is not passive space, but active temperature memory. Both arise only after temperature initiates rupture.

In conventional physics, motion is described as the displacement of matter through space over time, and geometry is treated as the static framework in which these displacements occur. Both are assumed to rely on time as a baseline coordinate. However, in the Unified Field Thermometry (UFT) model, this sequence is overturned. Time is not a precursor to motion or structure. Instead, temperature is the initiator—the primal scalar irregularity that precedes and generates both motion and geometry.

Temperature, in this model, is not classical heat or a function of kinetic energy among particles. It is the first asymmetry: a scalar differential emerging from zero-state equilibrium. It is a field condition, not an outcome. The etheric substrate, originally inert and unstructured, becomes active only when a thermal irregularity arises. That irregularity forces the ether to reorganize, producing motion and curvature in response to imbalance.

Motion is not spontaneous. It is an effect. It is what occurs when temperature violates symmetry and induces directional flow through structured vacuum channels. Energy begins to localize and move along these anisotropic pathways. There is no neutral motion in UFT—only reactive flow conditioned by thermal topology.

Geometry also does not preexist. It is constructed from the interactions between temperature gradients and field substrate. Where temperature varies across space, curvature emerges. The ether folds, twists, or stretches to accommodate differential pressures. These deformations give rise to structural coherence: the very fabric of geometry. Thus, dimensionality is not a container but an emergent characteristic. It is formed, not found.

The implications are foundational. Without temperature, nothing moves. Without motion, nothing differentiates. Without differentiation, there is no structure. Without structure, there is no geometry. And without geometry, there can be no time. UFT reframes the order of reality as:

Temperature → Motion → Geometry → Time.

This inversion reorients all physical assumptions. It suggests that causality begins not with events, but with conditions. Not with time, but with the rupture that permits change.

In summary, temperature is not merely a modifier of systems. It is their source. It initiates the response we call motion and sculpts the architecture we call geometry. Time is not the beginning. Temperature is.

This thermometric cosmology implies a new ontology. One in which the first distinction—the very seed of being—is a scalar variance in thermal potential. From this variance, all else unfolds.

8. 3 — Not Classical Heat, but Absolute Thermal Differential: The First Asymmetry

To speak of temperature in this model is not to speak of warmth or kinetic agitation. The temperature at the heart of UFT is a scalar rupture—a differential without a prior, the first inequality within an otherwise symmetrical field. It is not caused by particles colliding; it creates the very conditions by which particles might emerge.

This absolute thermal differential is the first asymmetry—the moment stillness divides. It is not diffusion, but disruption; not agitation, but genesis. It generates entropy by existing. It separates space into domains, assigning directionality where none existed. It is the scalar fingerprint of causation itself.

This foundational event is not subject to classical thermodynamics, which presuppose matter and time. Here, temperature is not an effect—it is the effective agent. It is what turns rest into rotation, symmetry into curvature, stasis into flow.

This is why the classical view of heat is insufficient. Classical heat describes energy transfer; this temperature precedes energy itself. It is the precondition for field differentiation. It causes light to exist by allowing contrast, by forcing rupture.

In the Unified Field Thermometry (UFT) framework, temperature must be severed from its classical interpretations. It is not heat as traditionally understood—not kinetic energy, not particle agitation, not a byproduct of combustion or conduction. UFT temperature is the scalar antecedent to all of these. It is not an emergent property of energy transfer but a pre-energetic condition. It does not arise from systems in motion; it initiates motion. This foundational redefinition is not semantic; it is ontological.

Classical thermodynamics defines heat as the transfer of energy due to temperature difference, yet always within the context of particles and matter. In UFT, there are no particles at origin, no energy trajectories, no spacetime backdrop. What exists at the boundary of nothing is a still field—perfectly symmetrical, inert, without vector. In this context, temperature is the first measurable rupture. It is the inaugural asymmetry in the scalar vacuum field: a pure differential without yet any transfer, mass, or motion. A breaking of stillness, not within matter, but within the substrate itself.

This absolute thermal differential is the event horizon of becoming. It marks the first deviation from equilibrium—a scalar imbalance that bifurcates potential and sets a boundary between what was inert and what will become active. Unlike classical heat, which requires a medium, this differential is medium-generating. It is not energy that spreads into pre-existing space. It is temperature that gives rise to space by deforming the uniform scalar field.

In this view, temperature is not energy per unit volume; it is the very metric that defines volume's emergence. The differential does not exist *within* space. Rather, space unfolds in response to the asymmetry. The radial expansion of field deformation—what will later be called curvature—arises in scalar response to the rupture. From this point onward, direction, flow, and structure are possible. But only because the symmetry was first broken.

This first asymmetry acts as a scalar gradient that cannot be reversed, only redistributed. It gives the field a preferential orientation, not in a Euclidean sense, but in a thermometric one. The asymmetry creates potential, and potential enables configuration. Thus, all later complexity—particles, forces, geometry, time itself—emerges from the memory of this initial imbalance.

Importantly, this is not an epistemic maneuver but an ontological postulate: that the existence of form presupposes an irreducible rupture in scalar unity. The UFT does not merely reclassify heat; it dethrones it. Heat, in its classical form, becomes a late-stage expression. The true origin is not heat but differential—a measurable inequality in the zero-field state that initiates phase distinction.

Therefore, the origin of the cosmos is not a bang, but a differential. Not a fireball, but an inequality. Not energy, but temperature-asymmetry without medium. This is the first cause. The primal pulse. The silent rupture that seeded all vibration, motion, and memory into a structureless void.

8. 4 — From Temperature Arises Structure, Motion, and Ultimately Curvature:

Gradients Initiate Structural Differentiation

Temperature does not simply disturb space—it organizes it. Within the UFT framework, thermal gradients act as the formative principle of physical reality. These gradients determine not only directionality, but structural form. Where temperature is scalar, gradients are directional. And it is this directionality that induces motion, carves structure, and induces curvature.

Channels in the vacuum lattice—conceptualized here as spatial conduits—begin to bend, swell, and fold in scalar response to these gradients. These deformations result in structured pathways: anisotropic vectors that define physical motion, directional time, and field behavior.

Structure, in this framework, is not accumulation but deformation. There is no assembling of particles—there is a thermal scaffolding that guides energy into form. Motion follows these paths as scalar reactions to field curvature. And curvature itself emerges not from mass, but from persistent thermal tension.

Thus, from temperature comes the ability to distinguish direction, to bind structure, to initiate waveforms. Gradients are not accidents—they are the architects of the field. Where they intensify, curvature increases; where they level, symmetry resumes. Without gradients, space is inert. With them, it becomes geometry.

The emergence of physical structure is not arbitrary, nor is it rooted in a priori geometries. In the Unified Field Thermometry (UFT) framework, structure is induced — not imposed — and its genesis can be traced to the earliest scalar irregularities of temperature. These thermal differentials are not classical gradients in a heat field; they are the first directional pulses in a previously undisturbed vacuum lattice. Temperature, as understood in UFT, becomes the primary agent of differentiation. It is the initial impulse that fractures symmetry and launches the field into organized complexity.

Where no gradient exists, no structure can form. It is only with the rise of thermal asymmetry that space begins to deform. The deformation is not random; it follows the contours of the imbalance. The vacuum lattice responds to this irregularity by folding, twisting, and realigning into channel-like pathways. These are not metaphoric structures, but literal anisotropic constraints within the field substrate. Spatial channels arise as thermodynamically sculpted conduits, etched by the magnitude and direction of temperature disparities. It is within these channels that matter will eventually localize and that energy will flow.

Structure, in this formulation, is geometric memory. Each bend in the ether marks a reaction to a prior imbalance. Each dilation encodes the thermodynamic stress that preceded it. It is not the mere presence of energy that determines the shape of space, but the relative differential of temperature across the field. High gradients induce sharp curvatures; low gradients maintain flat geometries. Thus, curvature becomes a readable index of thermal history.

Motion, likewise, does not precede structure. It is a consequence of it. Energy flows along the paths carved by prior temperature differentials. The so-called "laws of motion" are not fundamental edicts; they are statistical regularities emergent from these constrained channels. Velocity is dictated by gradient intensity. Directionality is constrained by the architecture of thermally-induced curvature.

It follows, then, that gravity is not a separate force but a sustained thermal effect. What appears as mass-induced curvature is, in the UFT model, a region of chronic temperature imbalance, maintained over time by localized asymmetry. In effect, the geometry of space is being continuously revised by the persistence of these gradients. Gravity is curvature is structure is imbalance.

This reordering reveals a profound conceptual shift: the universe is not built from matter interacting in space, but from temperature differentials sculpting the architecture that permits such interaction. Matter and energy are passengers of a deeper order—one that is thermometrically expressed and geometrically retained.

In summary, temperature initiates structure not as a side effect, but as a primal function. Structure is a differential response. Motion is a thermally-guided redistribution. Curvature is the memory of imbalance. Through this scalar chain of causality, the cosmos transitions from stillness to structure, from equilibrium to complexity, from void to form.

8. 5 — Time Is a Secondary Effect of Thermal Disequilibrium

Time does not govern reality; it is governed by thermal disequilibrium. Within UFT, time emerges only when temperature violates spatial symmetry and initiates field deformation. Each deformation records a thermodynamic event, leaving curvature behind as history. That history, read across scalar geometry, is what we interpret as time.

No symmetry, no time. No gradient, no rupture. Where equilibrium holds, no change occurs—and without change, no temporal index can exist. It is only when heat dislodges order, when flow distorts space, that time appears. And even then, not as a force or continuum, but as a visible accumulation of failure—of order succumbing to pressure.

Moments are not ticks—they are scars. They are memory layers in the field: geometric records of entropy's passage. These are spatial phenomena, not abstract coordinates. The past is that which is already ruptured. The future is that which remains under pressure but has yet to collapse. The present is the site of rupture, the place where temperature currently destabilizes the lattice.

Time, then, is not a cosmic clock but a thermal shadow. It is the outline of entropy's movement across structured space. It does not push matter—it trails it. It is a signature, not an actor.

In sum, temperature—not time—is the governing agent. It is not a result of motion, but its seed. Not a product of energy, but its initiator. All spacetime curvature, motion, and memory arise from this scalar origin—the pulse before time.

In the framework of Unified Field Thermometry (UFT), time is not elemental. It is not a primitive coordinate nor a governing continuum. It does not flow, exist independently, or serve as a universal backdrop against which events transpire. Rather, time is a derivative. It is a secondary phenomenon, emergent only from the dynamics of thermal disequilibrium. In this ontological reversal, time is not that by which we measure change. It is that which is produced when change occurs under asymmetric thermal conditions. It is not an input; it is a residue.

To understand time in this context is to abandon its traditional identity as a governing dimension. Instead, time is recognized as a structural consequence of temperature-induced rupture. Thermal disequilibrium initiates a process of structural deformation in the vacuum field—a deformation that reconfigures spatial geometry. These configurations, once deformed, do not reset. They persist. Their persistence becomes the archive of time. Each fold, each curvature, each channel displacement is a record of prior asymmetry. This record is what constitutes temporality.

The concept of thermal disequilibrium is pivotal. In a perfectly uniform thermal field—a vacuum of zero asymmetry—there is no motion, no structure, no curvature, and therefore, no time. Time cannot arise in the absence of thermal contrast. But once disequilibrium occurs—even infinitesimally—it propagates deformation across the field. The propagation is not instantaneous; it is patterned and sequential. That sequence is not time itself but the scaffold by which time becomes readable.

Entropy increases as the field reacts to these thermal imbalances. Each reaction, each deformation, leaves behind a scalar imprint. These imprints accumulate. Their accumulation forms the directionality, or arrow, of time. It is not that the field remembers in any cognitive sense. It is that its structure is permanently altered in a way that encodes thermodynamic history.

This history is not metaphorical. It is physically embedded in the scalar and vector properties of the field. It is visible in curvature shifts, measurable through channel displacement, and calculable through the universal spacetime constant $\Xi = E / (T \times V \times \pi r^3)$. In this formulation, temperature (T) is not merely correlated with time. It is its condition. Volume (V) and spatial curvature (πr^3) contextualize the geometry of deformation. Energy (E) is what is being distributed through the rupture. Together, they express the density of temporal becoming.

Thus, time is not sequential in the Newtonian sense, nor relativistic in the Einsteinian sense. It is scalar and structural, appearing only where symmetry fails. The present is the threshold where failure is active, where temperature gradients exceed equilibrium constraints and initiate new rupture. The past is not receding; it is cooling. The future is not advancing; it is latent—the domain of rupture yet unrealized.

In this model, clocks do not measure time. They respond to frequency changes induced by thermal differentials. Atomic oscillations are not universal temporal standards. They are local manifestations of scalar equilibrium responding to field asymmetry. Hence, different gravitational or thermal environments yield different local temporalities. Time is ecological. It varies with field structure.

Ultimately, time in UFT is a byproduct of heat refusing stillness. It is the visible, structured evidence of disequilibrium etched into the etheric substrate. It is not a substance. It is not an entity. It is a scar. A spatial consequence. A thermodynamic echo.

The Geometry of Rupture

Where the boundary is rupture and the trigger is temperature (Tempus Rupture)

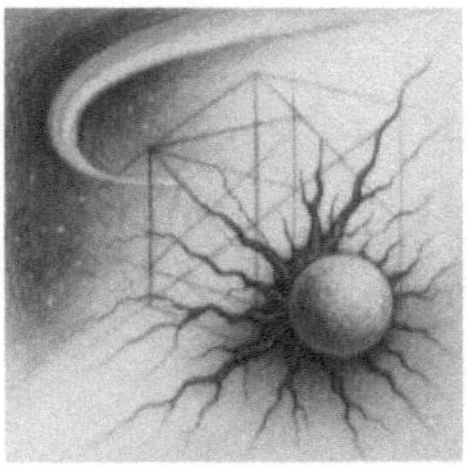

A graphite drawing depicts a dark, spherical core rupturing into branching fractures that spread into surrounding geometric scaffolding. The background features a swirling cosmic arc and subtle stars, symbolizing the emergence of structured space from implosive collapse. The image evokes the transition from thermal equilibrium to configured geometry—where time and curvature ignite.

KHRONUS

CHAPTER *NINE*
IV
Θ'THETA
ט TET
● ● ● ●

BOLON
NUEVE
9

THE GEOMETRY OF RUPTURE

9.1 — Ξ Begins to Operate Where Temperature Induces Configuration

The Universal Spacetime Constant, Ξ, does not govern a static cosmos. It activates only where structure begins—where temperature crosses from potential to configuration. In the Unified Field Thermometry (UFT) model, Ξ is not a universal background parameter but a scalar operator, a conditional expression that gains definition when thermal asymmetry generates spatial form. It is the signature of structure emerging from equilibrium collapse, the invariant quantity through which energy, curvature, and thermodynamic context resolve into geometry.

In classical models, constants such as c, G, or $\hbar$ are treated as universally active—ever-present, immutable. Ξ subverts this notion. It is not omnipresent; it is conditional. It *begins* to function only when temperature—the initiator—displaces symmetry and generates the first measurable differential. At the moment of configuration, when field conditions transmute from indeterminate potential to determinate form, Ξ enters.

This is not metaphysical suggestion. It is scalar causality. In the equation:

Ξ=ET×V×πr3Ξ = \frac{E}{T × V × πr^3}Ξ=T×V×πr3E

Ξ emerges as a ratio of energy to thermogeometric constraint. Energy alone cannot produce time, space, or curvature. Energy is pure potential. Only when constrained by temperature (T), spatial volume (V), and geometric curvature (πr^3) does energy become physically meaningful—only then is it shaped into behavior.

Ξ, therefore, is not the seed, but the birth certificate of configuration. It is the measure of how *much* structure is formed per unit of energetic input, once a system is ruptured from thermal equilibrium.

In the thermodynamic vacuum field, this activation point can be traced: prior to temperature-induced configuration, the field is an undifferentiated scalar medium. No geometry. No curvature. No direction. Ξ is undefined—not zero, but *inapplicable.* It does not vanish; it awaits. It is the mathematical parallel to a law that applies only when a boundary is crossed.

The boundary is rupture.

The trigger is temperature.

And the result is configuration—localized field structure that can then retain, transmit, or constrain energy. At that point, Ξ acquires measurable significance. It expresses the density and behavior of structured reality—not a theoretical realm, but the actual conditions through which physical phenomena unfold.

In this model, temperature is not one term among many—it is the gatekeeper. It determines when energy becomes bounded, when volume becomes meaningful, when curvature arises, and thus when Ξ becomes calculable. In a cold, inert system—at absolute equilibrium—Ξ has no bearing. The universe must first rupture into configuration before measurement begins.

This principle reframes the origin of physical law: not all constants are eternal. Some, like Ξ, are emergent—*they begin where structure begins.* This is the geometry of rupture: the transition from field potential to thermodynamically-bound structure. From there, Ξ governs the regime. It defines the proportionality between the energy housed and the spatial condition it is housed within. It anchors the configuration in a scalar relationship that cannot be broken without consequence.

Thus, Ξ begins where temperature induces configuration—not as decoration on the edifice of theory, but as its foundation.

9.2 — Implosion Precedes Expansion: Time Begins with Silence Collapsing into Detonation

There is no outward motion without inward collapse. In the Unified Field Thermometry (UFT) framework, expansion—cosmic or local—is never the origin. It is the consequence. What is seen as the birth of space is in fact the echo of its prior contraction. Every observable expansion begins with an implosive singularity: a moment where equilibrium fails so catastrophically that structure folds in on itself, compressing energy, breaking spatial rank, and initiating the conditions under which time becomes geometrically legible.

This is not metaphor. It is scalar logic.

Before any detonation—before curvature stretches, before geometry unfurls—there is a silence. This silence is not void, but **pre-rupture stillness**, a condition of unexpressed symmetry where all potential remains inward, suspended. In classical cosmology, this moment is obscured behind a Big Bang, a singular origin projected forward. But UFT reverses the lens: expansion is not the beginning of time. It is the response to implosive rupture. Time does not begin with movement outward, but with the first irreversible infolding—*the collapse that breaks the field's symmetry.*

This inward collapse—thermodynamic implosion—compresses localized volume under temperature differentials. Scalar pressure exceeds containment. The structured channels of the vacuum, once neutral, cannot hold. They buckle. As configuration density intensifies, curvature spikes. This creates not explosive force but **pre-explosive geometry**—the inward congealing of structure into rupture potential. The energy within does not yet radiate; it concentrates.

At this point, **Ξ remains dormant**—its denominator undefined, its scalar logic uninitiated. Then, under critical pressure, the implosive geometry reaches a rupture threshold. A singularity forms—not as infinite density, but as **maximal structural tension** within an asymmetric field. From this condition, detonation is inevitable.

Not because energy wills it, but because structure collapses beyond containment. Implosion gives way to release—detonation into curvature.

This detonation is not chaos. It is configuration release. It is **order emerging from structural pressure**, the scalar consequence of thermometric failure. The expansion that follows is not the original event, but the aftermath—the propagation of rupture through nested channels, the echo of implosion carved into space.

And this is where **time begins.**

Not as duration, not as continuity, but as rupture recurrence—a **quantized ignition**, the flicker that registers transformation. Each moment is not a tick but a scar, a record of collapse resolved through structured detonation. Implosion seeds curvature. Curvature manifests direction. Direction embeds entropy. Entropy defines time.

Therefore, time is not the prelude to explosion; it is the artifact of implosion. It begins **not with fire, but with collapse.** Not with noise, but with the silent breaking of symmetry—the irreversible transition from tension to release. In that moment, Ξ becomes real. Temperature has induced configuration. Implosion has birthed expansion. And time emerges—not as ruler of events, but as residue of their rupture.

9. 3 — Rupture Is Not Metaphor — It Is Measurable Transformation

In the Unified Field Thermometry (UFT) model, **rupture is not a poetic image**—it is a scalar event. A transition point. A measurable discontinuity in the structure of the vacuum field. Classical physics has long relied on smooth continuity: curves, trajectories, and conserved flows. But rupture interrupts this continuity. It introduces nonlinearity. It fractures the presumed stability of space and reveals that structure is not indefinitely elastic—it can fail. And when it fails, time begins.

Rupture, in UFT, is the moment when temperature gradients **overwhelm structural cohesion.** Within the channel-based etheric substrate, this failure is not uniform. It is localized, directional, and scalar—meaning it is governed not by force laws, but by threshold conditions: **a tipping point in curvature, density, and thermal asymmetry**. When these converge, the result is rupture.

This rupture is geometrically visible. It appears as a sudden distortion in spatial topology: sharp curvatures, implosive funneling, or elliptical seeding patterns. These are not approximations—they are **empirical fingerprints of past rupture events,** measurable through entropy shifts, photonic deflection, and anisotropic propagation. Just as seismic waves reveal the profile of an earthquake, spacetime curvature reveals the profile of rupture.

Rupture is also **thermodynamically irreversible.** Once the structural field has torn under asymmetry, it cannot revert to its prior state. Entropy increases. The rupture becomes a fixed record, and this record is what the observer perceives as a temporal event. Thus, time is inseparable from rupture—not conceptually, but physically. Without rupture, no time can be registered.

Importantly, rupture does not require motion. It requires **structural disequilibrium**. That is: a mismatch between thermal input and geometric capacity. When a given region of the field cannot accommodate the thermal load—because of insufficient volume, curvature, or pressure-dissipation routes—rupture occurs. It is **not a matter of scale.** Microscopic ruptures define quantum events. Macroscopic ones define gravitational domains. But the geometry is the same. The cause is the same. The metric of time is seeded identically in both.

Rupture also reframes causality. Traditional models attribute causation to chronological sequence. But in UFT, cause is spatial. Rupture triggers curvature. Curvature enables entropy flow. Entropy flow defines time. The arrow of time is therefore not an abstraction—it is the **irreversible spatialization of rupture.**

And rupture is not rare. It is recursive. The universe is not expanding from a singular event, but from nested ruptures, each generating localized spacetime domains. Each domain carries the signature of its rupture: frequency, amplitude, curvature, entropy state. The fabric of space is a **stratified archive of these ruptures**, and the map of their distribution is the map of time itself.

This model allows for prediction. Wherever field geometry approaches critical thresholds—of temperature, energy density, or curvature tension—rupture becomes statistically imminent. These thresholds are measurable. They define when and where time will ignite next.

Thus, rupture is not a metaphor. It is the scalar **mechanism of time emergence**, the observable process by which equilibrium fails and temporal geometry is born. And in UFT, it is the most fundamental transformation—**not an exception to order, but its very origin.**

9.4 — Transition from Equilibrium to Instability

In classical mechanics, systems tend toward equilibrium. Stability is the assumed norm, and disturbances are typically treated as perturbations—temporary deviations from an underlying order. But in Unified Field Thermometry (UFT), **equilibrium is the exception**, not the rule. The cosmos does not drift gently from order to disorder. It ruptures into instability—**and that rupture is the very act by which time begins.**

This chapter examines that **transition point**: the scalar threshold where a system crosses from thermodynamic balance into geometric instability. This crossing is not gradual. It is inflectional, sharp, and structurally consequential. It is here, in this transition, that structure gives way to curvature, and space becomes a record of its own failure.

In UFT, equilibrium is defined by **field symmetry**—a condition where temperature, curvature, and spatial configuration are isotropic and non-dynamic. No motion occurs. No entropy accumulates. Time, under this condition, is silent. It is not frozen—it is undefined. The system exists in a state of **thermal suspension.**

Instability, by contrast, is triggered when a **temperature gradient** becomes sufficient to distort spatial channels. This is not merely an energetic excess. It is a structural crisis: a point at which **the underlying lattice of vacuum channels can no longer absorb, redistribute, or regulate the energy imposed upon it.** When that threshold is passed, the system collapses—not into destruction, but into differentiation.

This is the pivot—**the scalar event** in which spatial homogeneity fractures into anisotropy. The field becomes directional. Energy flows become guided, not diffuse. Entropy no longer disperses uniformly; it localizes, moves, bends space. The ether ceases to be neutral. It becomes **active**—structured, curved, and temporal.

The transition from equilibrium to instability thus redefines motion itself. In classical physics, motion presumes time. But in UFT, **motion is not a function of time—it is the fingerprint of instability.** Once symmetry fails, energy moves. Once energy moves, structure deforms. Once structure deforms, curvature registers. And once curvature registers, **time is measurable.**

The moment of instability can be identified through several geometric and thermodynamic signatures:

Localized increase in scalar curvature, indicating spatial tension.

Formation of elliptical channel deformations, precursors to directional motion.

Increase in entropy vectors, pointing away from rupture origins.

Reconfiguration of the Ξ constant, where $E / (T \times V \times \pi r^3)$ undergoes a spike or inflection.

Each of these indicators reflects the same root transformation: the **birth of temporal topology** from thermal imbalance.

This transition is not theoretical. It occurs continuously—at quantum scales in field fluctuations, at astrophysical scales in stellar collapse, and at cosmogenic scales in galactic seeding. The cosmos is not an equilibrium machine. It is an instability engine. And that instability is not failure—it is formation.

UFT therefore teaches that **equilibrium is not the goal of the universe, but its raw material.** Instability is not decay—it is structure. And the moment of transition is not disorder, but the initiation of a new order, sculpted by rupture and sealed in entropy.

Time, in this frame, is not the child of order—but the signal of its defeat. It is not continuity—but instability crystallized.

9. 5 — Time as a Rupture Rate in a Bounded System

Time is not a continuum. It is not a flowing ribbon upon which events are threaded. In the Unified Field Thermometry (UFT) model, time is a **scalar consequence**—a quantifiable **rupture rate** across a **bounded, structured field.** It does not stretch or loop or tick; it fractures. It is not a river—it is a pattern of detonations.

In classical physics, systems are assumed to evolve smoothly over an independent time axis. Time is treated as an external framework: absolute, continuous, and infinitely divisible. But in UFT, time **does not exist until rupture occurs.** And rupture is never infinite. It is always **bounded**—physically, spatially, and thermodynamically.

A bounded system, by definition, has **finite structure**—volume (V), curvature (πr^3), and temperature (T). Within such a system, time only emerges when these internal conditions reach thresholds that cannot be internally compensated. When energy exceeds the capacity of structural channels to absorb or redistribute, the system ruptures. That rupture is not chaotic—it is **scalar**, and its **rate of recurrence** becomes the functional identity of time.

Thus, **time is the metric of instability per spatial unit:**

Time = d(rupture events) / d(volume × entropy gradient)

This is equivalent to saying that time is **the rate at which a bounded system exceeds its structural thresholds** and reorganizes geometrically.

Each rupture is thermometrically defined by the Ξ constant:

$\Xi = E / (T \times V \times \pi r^3)$

As temperature gradients steepen within confined geometry, Ξ increases. Each pulse or spike in Ξ marks a **local rupture event**, and the **frequency of those pulses** defines a scalar temporality. Not the illusion of flow, but the **rhythmic succession of structural instability**. In short:

High rupture rate = fast time

Low rupture rate = slow time

Zero rupture = no time

This conception also clarifies gravitational time dilation. Near dense masses, curvature is steep, volume compressed, and temperature conditions intensified. The rupture rate is altered—not by time itself slowing, but by the **field's internal structure** modulating how often rupture conditions are met. That modulation is time.

The bounded system can range from quantum vacua to stellar interiors. But across all domains, **it is not motion that measures time—it is rupture that defines it.** Time is not fundamental—it is **episodic.** Every moment is a scalar scar. Every duration is an **interval between structural reorganizations.**

This interpretation dissolves the paradox of time travel. One cannot "move" in time, because time is not a medium. One can only measure **how many ruptures have occurred within a given scalar system**—a thermodynamic count of failure, not a linear path.

In summary:

Time is not an external framework.

Time is a scalar metric of rupture within spatially bounded thermodynamic fields. Ξ governs the local rupture logic; rupture rate defines temporal experience.

No rupture = no change = no time.

In this light, the universe does not age. It ruptures. And its age is nothing more than the **total record of ruptures** across the structured ether.

Time, then, is not merely a consequence of heat. It is the **fractured grammar** of a cosmos that burns.

CHAPTER *TEN*
X
I'OTA
י YOD

LA-JUN
DIEZ
10

TIME DIES WHERE HEAT FAILS (INTERLUDE)

10.1 — No Temperature Gradient → No Entropy → No Time

Time requires asymmetry. It does not exist in equilibrium. In the Unified Field Thermometry (UFT) model, the progression of time is strictly contingent on the presence of a temperature gradient. Without differential thermal conditions, there can be no entropic shift. Without entropy, there can be no irreversible deformation. And without irreversible deformation, there is no metric for temporal distinction.

In a perfectly uniform thermal field, energy neither flows nor organizes. Structure remains inert, curvature remains undeformed, and the etheric channel lattice exhibits no rupture. This is the condition of temporal nullity—a spatial zone where time does not pass because nothing occurs to mark its passage.

In the Unified Field Thermometry (UFT) framework, time does not exist as an independent background condition—it must be triggered into being. That trigger is not energy in isolation, nor matter per se, but thermal asymmetry. The chapter's opening axiom encapsulates this hierarchy: *no temperature gradient → no entropy → no time.* This is not a rhetorical progression but a scalar law. Each condition is causally dependent on the previous. Temperature gradients are not the consequence of motion—they are its origin. From this asymmetry, entropy arises. And from entropy, the residual effect we call time becomes measurable.

10.1 — No Temperature Gradient ↓ No Entropy ↓ No Time

To understand why temperature is foundational, one must first reject the classical misconception of it as merely "heat." In UFT, temperature is scalar distortion within the vacuum substrate—a field imbalance. Without such gradients, all field vectors remain isotropic; structure remains inert; no direction can be resolved, no motion initiated, no difference recorded. In such a perfectly symmetric condition, there is no possibility of entropy increase—because entropy presupposes distinguishable states. If every region is the same as every other, no disorder can grow, and no informational sequence can be formed.

This absence of disorder is not stasis; it is non-being. The field remains unruptured. Without temperature differential, space cannot be curved, structure cannot be inscribed, and energy cannot express trajectory. The etheric substrate remains flat—not geometrically, but informationally: no hierarchy, no fold, no pathway, no cause. In such a configuration, the concept of time becomes inapplicable. There is nothing to measure, no transitions to record. Time is not slow or halted—it is undefined.

Entropy is the first derivative of temperature imbalance. Where gradients exist, entropy emerges as the rate of disorder propagation through the substrate. But if temperature is uniform across the system—if there is no gradient—then entropy has no domain over which to increase. The Second Law of Thermodynamics becomes trivially true: nothing changes because nothing can. The vacuum remains unstructured, the field remains unaltered, and the arrow of time never forms.

In UFT, this insight transforms the metaphysical status of time. Time is not the stage, nor the origin. It is the index of rupture—a scalar byproduct of asymmetric thermometric activity. It does not exist in systems at thermal equilibrium, because no gradients exist to destabilize the structure. And since no structure is deformed, no record is left. Time requires memory—not as consciousness, but as curvature, scar, and trace.

This also reframes black holes and absolute zero environments. In conventional physics, these are extremes. In UFT, they are ontological thresholds. A perfect black hole core—a region where thermal differentials collapse to nullity—would become temporally blind. Likewise, at absolute zero, where temperature ceases to act, the substrate reverts to non-time. Not a stillness in duration, but a return to precondition.

Therefore, time is not present wherever there is temperature. It is present only where there is **differential temperature**—that is, a gradient. Flat thermometric fields are temporally invisible. It is the fold, not the presence, of heat that creates history. In UFT, this singular insight becomes the bedrock of cosmological reordering. Not mass, not energy, not geometry—but *gradient*—is the generator of time.

Thus, the arrow of time is not merely aligned with entropy; it is etched by it. But entropy, in turn, is etched only by the uneven hand of temperature. And temperature arises only as rupture. Where there is no rupture, there is no entropy. Where there is no entropy, there is no time. Where there is no gradient, nothing ever happens—and nothing ever has.

10.2 — Flat Structure Implies Temporal Stillness

Geometry is not neutral. It responds to imbalance. Where the field geometry is flat—absent of stress, curvature, or flow—there exists no record of instability, and hence, no time. A flat field represents a frozen state of the cosmos: no rupture, no scar, no succession.

This is not a hypothetical condition. In deep cosmic voids or stabilized laboratory vacua approaching absolute zero, time functionally slows to a halt. No curvature evolves, no entropy accumulates, and no geometric memory is formed. The system may persist, but it does not age.

Flatness, therefore, is not the absence of activity—it is the absence of thermometric causality. It is the thermal silence that precedes ignition, or the stillness that remains when rupture ceases.

10.3 — Light and Entropy Are the Only Lingering Witnesses of Past Ruptures

Even in the absence of ongoing rupture, the past does not vanish. It echoes. Light and entropy act as the remaining traces of prior rupture events. Photons traverse structured space as vectors of entropic memory—each wavefront a timestamp of curvature crossed, each path a legacy of thermal imbalance.

Entropy, too, is not instantaneous. It accumulates. Its distribution maps the scalar history of system failure and geometric reconfiguration. These are not abstract concepts: they are embedded in field density, in radiation spectra, in the very shape of space itself.

In still systems, what remains is not time itself but its residue. The glow of ancient heat. The spread of imbalance that once was. Time dies where heat fails, but its ghost moves at light speed.

In the Unified Field Thermometry (UFT) framework, spatial configuration governs the presence or absence of time. When spatial channels remain undisturbed—perfectly balanced, isotropic, and flat—no directional energy flow occurs. Without gradients to induce deformation, and without curvature to channel oscillatory behavior, the field persists in a state of stasis. This is not merely the absence of change; it is the preclusion of time itself. Temporal progression, in this sense, is not suspended—it is never initiated.

Flat structure implies a total equilibrium of energy distribution. In this geometrically neutral field, no preferential pathways exist for energy transfer or structural reconfiguration. Spatial curvature, which in UFT encodes the historical residue of entropy-driven deformation, is absent. Without deformation, there is no entropy vector. Without entropy, there is no causal flow. And without causal flow, there is no metric progression that could be interpreted as time. The field, though existent, is temporally dormant.

This stillness is not theoretical but definable: it is the zeroth condition, the null topology from which asymmetry and structure may later emerge. The implication is profound—time cannot exist in a system that lacks thermometric differentiation. A perfectly flat structure does not merely avoid the passage of time; it does not admit the ontology required for time to exist.

In such a state, the speed of light is undefined, not because light cannot propagate, but because there are no pathways through which to propagate. Gravitation is non-expressive, as there are no mass-energy differentials to deform the channel architecture. The system is not in stasis due to internal resistance or entropic decay—it is silent because no condition exists to provoke its rupture. Thus, temporal stillness is not a passive state, but a precondition of absolute structural uniformity.

Moreover, this concept challenges the Newtonian and relativistic assumption that time is universally present, merely modulated by relativistic effects. In UFT, time does not scale—it ignites. It arises only where structural disequilibrium breaks the silence of geometric flatness. Therefore, a flat spatial structure is not simply temporally inert. Factually it is a witness to time's non-material existence. It is a domain of zero entropy, zero directionality, and thus zero becoming.

This principle plays a vital role in cosmological interpretation. The pre-expansion vacuum—often imagined as a 'timeless' void—was not merely devoid of events; it lacked the structural preconditions for time to arise. Temporal stillness is not an emptiness awaiting activation. It is a strict condition: a geometrically null, thermally symmetric substrate that prohibits the very birth of motion and memory. From such stillness, only a rupture—driven by thermal imbalance—can initiate the transition into spacetime dynamics.

Thus, in the UFT model, flatness is not a geometric triviality. It is an ontological firewall against time. No temperature gradient, no curvature. No curvature, no entropy. No entropy, no time. Flat structure is not pre-time—it is anti-time, the geometrically encoded rejection of sequence, causality, and change.

10.4 — Time Nullification Where Temperature Gradients Approach Zero

As temperature gradients approach zero, the energetic tension that sustains temporal flow diminishes. In UFT, the scalar evolution of time is directly proportional to thermal asymmetry—meaning, time is sustained only by a differential in local temperature. As the gradient flattens, directional flow ceases, and with it, the ability of the field to inscribe entropy through structural deformation vanishes.

This is not a metaphysical suspension of time—it is a scalar termination of temporal causality. The lattice that underlies space, absent directional flux, lacks the condition to generate curvature or channel rupture. With no entropy production, no irreversible deformation occurs. The field, still present, no longer bears witness to change. In such domains, time is not dilated—it is extinguished.

In this formulation, null-time regions could exist at the thermal boundaries of the universe or in artificially stabilized environments approaching absolute equilibrium. This state is not merely cold—it is geometrically and causally sealed. Time, which elsewhere functions as a cascade of thermal scars across spatial channels, is here nullified by perfect stasis.

Observable effects may include phase-invariant regions or quantum systems locked in coherence, where probabilistic behavior fails to evolve. Unlike classical time-stopping scenarios in relativity (such as inside event horizons), this UFT-based nullification is thermodynamic and scalar—not geometric and relativistic. It results from the internal structure of the vacuum being freed from deformation, not from relativistic collapse.

Thus, time nullification is not catastrophic; it is a terminal calm—a field returned to silence. It reflects not decay, but the absence of thermal voice. In approaching zero gradient, the universe, locally or universally, approaches not death—but mute perfection.

This condition is not merely theoretical. It defines the deep core of gravitational wells, the interiors of perfect Bose-Einstein condensates, and the silent ether beyond thermodynamic activation. Here, Ξ collapses toward an undefined state. The denominator dominates: $T \rightarrow 0$, $V \rightarrow$ constant, $\pi r^3 \rightarrow$ finite. Time cannot emerge when Ξ becomes immeasurable.

Time is thus not broken in such regions. It is unborn. Like unstruck chords, its potential exists, but without asymmetry, it never sounds.

10.5 — Thermodynamic Flattening = Cessation of Entropy = Cessation of Time

In the Unified Field Thermometry model, the progression of time is inseparable from thermodynamic disequilibrium. Time does not advance because of clocks—it advances because entropy increases. But entropy itself only advances where temperature gradients exist. Thus, where thermal gradients are extinguished, entropy plateaus. And when entropy ceases to evolve, time collapses into nullity.

Thermodynamic flattening represents this ultimate state. It is the condition where all structural deformation has ceased. No curvature, no stress, no directional flow of energy. Channels remain in symmetrical equilibrium. Without stress on the field, no spatial memory is written. No new geometry forms. No history is etched.

In such a scenario, the etheric substrate becomes temporally inert. Events lose sequence. Processes halt. There is no thermodynamic signature left to differentiate one moment from another. Time is not merely paused—it is erased as an operative dimension. The temporal arrow, once defined by the advance of disorder, disintegrates when disorder becomes impossible.

This is the asymptotic fate of thermal dynamics. Whether locally or cosmically, when systems approach this flatline, they exit temporality. They enter silence—not of absence, but of perfect symmetry. The field exists, but it no longer changes. Therefore, it no longer tells time.

This reframes time not as an ever-running river, but as a localized response to energetic imbalance. Once that imbalance is resolved—once every gradient dissipates—the river evaporates. What remains is the unbroken mirror of thermodynamic stillness: existence without story, space without sequence, reality without time.

Entropy is the arrow. Time is the path. But when the thermodynamic field flattens, the arrow drops.

Flattening is not a collapse. It is a stabilization. When energy disperses perfectly, when gradients dissolve into uniform equilibrium, there is nothing left to drive asymmetry. The system no longer requires change, and so time no longer has a medium to traverse.

This is the final state of all closed systems. Not death, but silence. A system in which no moment differs from the next. Where all structure is equalized and the field has nothing left to express.

In this state, time does not stretch on. It vanishes. And only its residues—light and curvature—remain as the fossil record of heat once burning.

Appendix A — Dimensional Analysis of Ξ

The Universal Spacetime Constant: $\xi = E / (T \times V \times \pi r^3)$

Where:

- E is energy [$kg \cdot m^2/s^2$]
- T is temperature [K]
- V is volume [m^3]
- πr^3 encodes spherical spatial curvature [m^3]

Combined Denominator Dimensions: [$T \times m^3 \times m^3$] = [$K \times m^6$]

ξ Units: [$kg \cdot m^2/s^2$] / [$K \cdot m^6$] = [$kg / (K \cdot m^4 \cdot s^2)$]

Interpretation: ξ is a scalar density of energy per unit thermal-structural field volume. It defines the energy concentration required to initiate a curvature event at a given temperature and spatial configuration.

Implications:

- ξ is invariant within a given curvature regime.
- It allows field transformations to be modeled across relativistic and quantum domains.
- It ensures dimensional compatibility with both thermodynamic gradients and gravitational curvature.

Cross-Domain Relevance:

- Thermodynamics: Regulates entropy formation.
- Quantum Field Theory: Governs vacuum excitation thresholds.
- Relativity: Correlates mass-energy distribution with spacetime curvature.

The Universal Spacetime Constant serves as the grounding scalar through which rupture can be geometrized, entropy registered, and time born.

Implosion Expansion Dynamics in thermometric fields

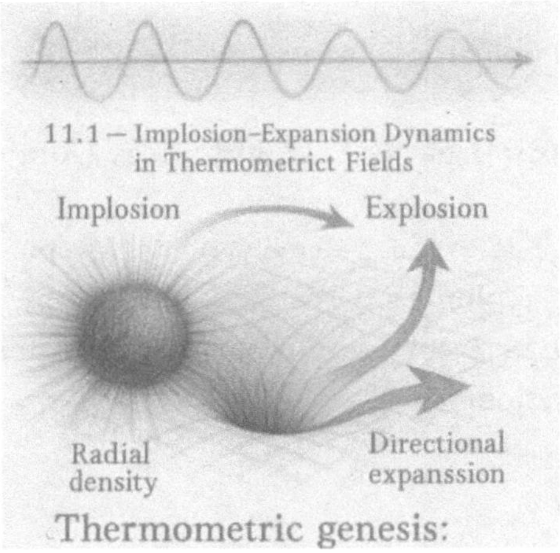

The graphite sketch depicts the dynamics of implosion and expansion within thermometric fields. A dense, shaded sphere on the left represents radial density and implosion, with arrows curving outward to show directional expansion and explosion. Above, a wave-like frequency line symbolizes time as structured oscillation. The background features a subtle grid, suggesting spatial channels and field structure, with clean labels guiding the viewer through the cosmic process.

CHAPTER ELEVEN
XI
ια' IOTA-ALPHA
א"י YOD-ALEPH

BULUK
ONCE
11

TIME AS A STRUCTURED FREQUENCY

11.1 — Implosion–Expansion Dynamics in Thermometric Fields

Thermometric Genesis: The Primordial Sequence

In the framework of Unified Field Thermometry (UFT), cosmic dynamics are not driven by abstract inflationary models or arbitrary forces, but by thermal asymmetries structured through scalar field behavior. The genesis of the cosmos is not characterized by a linear "big bang," but rather by a thermometric sequence: first implosion, then regulated expansion. These are not opposite actions, but conjugate phases of a scalar field's response to temperature imbalance within a structured vacuum.

Implosion marks the primary thermal contraction of inert matter—still timeless and static—before any directional flow or structural differentiation. It is the drawing inward of thermometric potential, a compression of scalar uniformity into a point of field intensity. This is not mass collapse in the relativistic sense, but a gathering of spatial channel symmetry—a condensation of undisturbed etheric paths. Implosion is the field's preparatory response to an imposed temperature asymmetry.

Only after critical deformation thresholds are crossed does expansion occur. Expansion is not triggered by energy overflow, but by field restructuring under pressure gradients. It is the spatial re-articulation of curvature, the rapid realignment of channels previously compressed. From implosion arises pressure. From pressure arises rupture. From rupture arises geometrized expansion—not as explosion, but as systemic field expression, where structure unwinds and entropy scales into motion.

The Implosion-Expansion Cycle as Scalar Law

This duality is not incidental but universal. The implosion-expansion cycle is intrinsic to thermometric fields because it expresses the minimum condition for structure to form. No geometry appears unless something first collapses under scalar stress. No motion is initiated unless that collapse finds directional symmetry-breaking.

In this cycle:

- **Implosion** compresses entropy into high-density curvature packets.

- **Rupture** signals a scalar phase transition—the point where the vacuum can no longer maintain curvature stability.
- **Expansion** distributes that curvature outward along temperature-biased channels.

The rupture event marks the critical moment when spacetime begins to behave thermodynamically. Time, in this context, is born not at "bang," but at rupture—the transfer point between thermometric implosion and geometrical expansion. It is not explosion that creates time, but the field's scalar instability when resisting further compression. This is where Ξ—the Universal Spacetime Constant—begins to operate as a measurable regulator.

Thermal Fields Precede Kinetic Force

The conventional kinetic view sees the universe expanding due to inertia or initial motion. But UFT reverses this: space expands because the field itself is under thermometric reconfiguration. The vacuum conducts expansion like a hydraulic fluid under asymmetric thermal compression. Implosion does not launch particles but restructures geometry—allowing matter and motion to follow paths already etched by scalar instability.

Expansion is thus not the movement of matter outward, but the propagation of structural deformation through the channel matrix. In this model, the velocity of expansion is not constant, but conditioned by temperature gradient, entropy flux, and curvature modulation. Local differences in curvature density govern how far, how fast, and in what configuration the expansion proceeds. It is not space that "grows," but rupture that migrates through a medium pre-structured by implosive conditions.

Implications for Observational Cosmology

This view reframes cosmic background radiation not as remnant heat from an explosion, but as the observable imprint of a large-scale rupture event—an energetic echo of field instability. Anisotropies in this background are not statistical irregularities but directional residues of implosive resistance—scars of thermometric memory embedded in the vacuum's initial phase transition.

Likewise, redshift does not necessarily imply motion through space, but the thermometric attenuation of coherent oscillations as channels stretch under entropy gradients. Expansion in UFT is not a function of velocity but a function of curvature release. Light shifts not because galaxies flee, but because space itself reconfigures in response to temperature differential and channel strain.

Summary

In *Cosmic Thermometry*, implosion and expansion are not events, but sequential field states governed by scalar thermodynamic law. Implosion collects and intensifies the field; rupture registers instability; expansion unfolds structure. This cycle is not explosive—it is thermometric. It follows not arbitrary kinematic rules, but the logic of curvature release under temperature stress.

This perspective dissolves the classical myth of a single explosive beginning and replaces it with a structured, continuous process of field adaptation, scalar rupture, and directed spatial articulation. The universe does not begin with motion. It begins with imbalance. And from that imbalance, everything proceeds.

11.2 — Singularities as Localized High-Tension Packets in Spatial Channels

Redefining the Singular: From Infinite Collapse to Scalar Localization

In conventional relativistic physics, a singularity is understood as a point of infinite density, infinite curvature, and undefined geometry—a breakdown in the equations of General Relativity. UFT rejects the need for infinite values and reframes singularities as **localized high-tension packets** within the structured spatial channel network. These are not metaphysical anomalies but scalar field concentrations—zones of extreme curvature and thermometric compression that mark where structure cannot remain uniform.

Rather than "tears" in spacetime, singularities in UFT are the extreme limit of channel elasticity. As thermal gradients concentrate in a confined region, the surrounding channels absorb more stress than their structural coherence can handle. This creates a self-compacting node, a dense vortex of channel absorption and curvature alignment. These nodes are not infinitely small, but **discretely bounded**, defined by the capacity of the field to store scalar stress without rupture.

This reinterpretation preserves mathematical rigor while offering a physically intelligible alternative: singularities are **field-bound condensates**, not infinite points. They serve as spacetime's most intense memory of implosion, compression, and scalar asymmetry.

Thermometric Signature of a Singularity

Every singularity within UFT is characterized by a **radially increasing temperature differential** that approaches critical thresholds. As one nears the singular node, the local temperature gradient becomes steep enough to cause:

- Channel collapse (localized absorption of field vectors),
- Radial curvature intensification (geometric folding of space),
- Time dilation (due to increased entropy compaction),
- Radiation deflection or absorption (electromagnetic behavior altered by curvature density).

This high-tension environment transforms ordinary field behavior. Entropy gradients steepen, directional symmetry breaks, and motion ceases to be linear. At the threshold, **motion becomes curvature**, and **curvature becomes a scalar record of thermal compression.**

Such behavior does not require the violation of known laws but provides an expanded thermometric geometry through which known phenomena—gravitational collapse, event horizons, and high-energy emissions—can be modeled without conceptual breakdown.

Spatial Channels Under Stress

Spatial channels, as described in UFT, are not passive carriers of mass-energy—they are **elastic conduits of structural information.** Under typical conditions, they maintain equilibrium by balancing absorption and emission processes. However, when thermal differentials rise beyond channel absorption capacity, **channel integrity collapses inward.**

This inward collapse is not a singular point but a **scalar process:**

- At first, channel density increases.
- Then, orientation narrows—channels align radially toward the heat source.
- Finally, a rupture occurs—not in material but in configuration, resulting in a high-curvature, low-emission packet: a thermometric singularity.

From the outside, this packet appears as a mass-density anomaly—hence the classical association with black holes. But under UFT, what is observed gravitationally is in fact a **deformed field configuration**, a thermally-induced compression scar, not a point of infinite energy.

Implications for Black Holes and Compact Objects

Within this framework, black holes are not collapsed stars per se, but **scalar compressions of field structure**—implosive reconfigurations of spatial channels under extreme thermometric strain. Hawking radiation, in this model, is interpreted not as quantum tunneling but as **residual emission** from partially restored channel emission at the rupture boundary. Event horizons mark the outer limit of scalar readability—where curvature becomes steep enough that information cannot escape outward oscillation.

This perspective also reinterprets neutron stars, pulsars, and other compact bodies as **thermometric stabilizers**—localized field configurations where structural collapse stops short of rupture, maintaining equilibrium through sustained high-channel tension.

Thermometric Containment and Rupture Thresholds

Unlike classical singularities, which imply a one-way collapse, UFT posits that singularities are **reversible in principle**—given sufficient redistribution of thermal stress. The containment of high-temperature differentials within bounded spatial curvature implies that under certain conditions (e.g., entropy dissipation, external gradient disruption), the singularity can **resolve**, releasing its curvature through radiative expansion.

This reversibility is fundamental to understanding **cosmic cycles**, where implosion precedes expansion, and where each rupture becomes a site for new curvature propagation. Singularities, therefore, are not terminal—they are **pre-expansion seeds**, condensates of potential structural differentiation.

Summary

Singularities in *Cosmic Thermometry* are not metaphysical infinities or theoretical artifacts. They are **localized thermometric compressions**, expressed as curvature packets within the etheric channel lattice. Their behavior is governed by temperature gradients, not by absolute collapse. They store entropy, alter curvature, modulate radiation, and—crucially—initiate the next phase of spatial expansion.

Under UFT, the singularity is redefined: no longer the end of physics, but the **pivot point** where implosion

11.3 — Implosion → Radial Density; Explosion → Directional Expansion

The Scalar Directionality of Collapse and Emergence

In conventional cosmological narratives, the universe begins with a singular "bang"—a moment of expansion without a defined precursor. In the UFT framework, this picture is reversed and reframed: **implosion precedes expansion.** This inversion is not metaphorical but scalar—the field contracts thermometrically before it radiates structurally. Implosion is not destruction; it is **radial condensation,** a scalar packing of energy into increasingly dense, coherent curvature. From this localized stress emerges the geometry of expansion.

Implosion concentrates, organizes, and embeds energy within high-density channel configurations. Explosion, by contrast, is not a violent outburst but a **structurally governed re-expansion**—an emission of curvature via reorganized spatial channels. One compacts, the other diffuses. One draws energy inward along radial vectors; the other distributes energy outward through differentiated channels.

The asymmetry between implosion and explosion is not directional alone—it is **thermodynamic.** Implosion increases entropy density (per unit volume), while explosion distributes entropy across emergent curvature. One embeds time in structure; the other releases time through motion.

Implosion as Radial Density Increase

Implosion in UFT is not gravitational collapse in the Newtonian sense. It is a **scalar intensification:** a thermometric process that compacts entropy and mass-energy into smaller and smaller volumes via radial channel absorption. This process causes: —

- Local increase in spatial channel density.
- Collapse of emission behavior (electromagnetic silence).
- Conversion of motion into curvature (kinetic arrest).
- Extreme thermal gradient compression (temperature spike).

Radial vectors converge toward a central locus, folding spatial orientation inward. Time slows as entropy condenses. The field becomes **temporally loaded**—poised to rupture outward once critical thresholds are exceeded.

Implosion, then, is the **geometry of scalar loading.** It defines the moment where energy can no longer remain silent, where temperature differentials saturate the field, and where structure ceases to absorb without deforming.

Explosion as Directional Expansion

Explosion is not simply the inverse of implosion. It is **structured release.** Rather than a chaotic scattering, it unfolds as **directional expansion**, governed by the scalar configuration laid down during the implosive phase. As channels under stress regain emission behavior, they propagate energy along coherent trajectories. What appears as random dispersal is actually **patterned spatial unbinding**:

- Channels realign outward, reducing curvature locally.
- Stored thermal pressure resolves into vector motion.
- Structural scars become pathways for radiation.
- Time accelerates as entropy distributes.

Explosion is thus a phase transition from thermometric saturation to **radiative coherence.** The universe does not explode because it was born; it explodes because **it could no longer remain compressed.**

Entropy Vector Inversion

In UFT, entropy is not merely a statistical count of microstates. It is a directional field vector, an imprint of thermodynamic asymmetry. During implosion, entropy vectors converge—embedding memory into curvature. During explosion, they invert—propagating memory through radiation. This inversion creates a scalar timestamp, a thermodynamic arrow that defines the irreversibility of expansion.

This provides a new foundation for the arrow of time: it is not an arbitrary orientation but a geometric trace of entropy's directional inversion. Implosion writes time into curvature. Explosion reads time from that structure and emits it into space.

Transition Dynamics: From Scalar Compression to Field Emission

The moment of transition between implosion and explosion is not instantaneous. It is mediated by a threshold: the rupture point, where scalar stress exceeds the structural tolerance of spatial channels. This rupture is not destruction—it is emission through failure:

- The field yields.
- Channels reconfigure.
- Radiation begins.

This rupture redefines "beginning" not as creation from nothing, but as **configuration from saturation.** The Big Bang, in this model, is not the birth of matter—it is the **emergence of channel reorganization** following implosive scalar saturation.

Implications for Cosmology

This framework radically reinterprets cosmic evolution:

- The universe did not emerge from an explosion alone—it first collapsed into high-order scalar density.
- Time began not at the point of expansion, but at the moment implosion could no longer be contained.
- Radiation (light) is not the first thing; it is the **first released memory** of thermal asymmetry.

This offers new explanations for:

- Cosmic background uniformity (implosive compression precedes emission).
- Horizon and flatness problems (pre-expansion organization).
- Dark energy dynamics (as residual scalar field tension from incomplete expansion).

Summary

Implosion and explosion are not opposites—they are sequential thermometric states within the UFT model. One defines scalar concentration, the other scalar expression. Implosion compacts curvature through radial density. Explosion releases curvature through directional expansion. Together, they form a thermodynamic cycle: compression → rupture → propagation.

This is the heartbeat of the cosmos—not random, not violent, but structurally inevitable. A scalar breath: inward tension, outward revelation. Time begins in the rupture between the two.

11. 4 — Ether Conducts Rupture Like Thermometric Hydraulics

The Etheric Substrate: Not Passive, but Responsive

In the Unified Field Thermometry (UFT) framework, "ether" refers not to a discarded nineteenth-century concept, but to a structured vacuum medium composed of interconnected spatial channels. These channels, while invisible to classical detection, serve as the architecture of physical reality—conduits for mass-energy, entropy flow, vibrational resonance, and curvature deformation.

Within this architecture, rupture is not an abstract event; it is a hydraulically transmitted thermodynamic reaction. The ether behaves like a fluid under pressure—not in terms of molecular agitation, but as a scalar-responsive lattice capable of transmitting deformation, resisting differential tension, and redirecting thermal imbalance along geometric vectors. Rupture does not simply happen in the ether; it *propagates* through it like pressure through a complex hydraulic system.

Temperature Gradient as Scalar Pressure

In traditional fluid systems, pressure results in flow. In UFT, temperature differentials function analogously to pressure gradients. A sharp thermal discontinuity—across a boundary or localized field distortion—induces scalar tension within the etheric lattice. This tension is not immediately visible but thermodynamically real, embedded in the stress geometry of the spatial channels themselves.

As the thermal gradient increases beyond a critical threshold:

- The ether resists—temporarily absorbing the tension.
- The channels stretch, bend, and warp—altering curvature locally.
- If the gradient becomes unsustainable, **rupture propagates.**

This rupture is scalar—not a break in matter, but a release of structural coherence. The ether yields not by tearing but by reorganizing: **thermal stress flows through the medium**, deforming its channel configuration in measurable, directional ways.

Rupture Propagation is Not Explosive, But Directed

Rupture in the ether does not mimic explosions in air or matter. It does not blast—it **streams.** Like a hydraulic impulse that travels through fluid lines, rupture follows the line of least thermometric resistance:

- High-entropy zones push into low-entropy domains.
- Channel pathways guide the rupture's movement.
- Temperature fields determine the rupture's velocity and direction.

Thus, **ether conducts rupture** in patterns traceable by thermal geometry, not by classical force dynamics. This reveals a scalar law: **energy flows not where it is "pulled" but where channel tension permits emission.** The ether does not act; it configures response.

Thermometric Hydraulics and the Logic of Cosmos

If we consider cosmic behavior—star formation, black hole jets, galaxy filamentation—it follows that much of this can be reinterpreted as **thermometric hydraulic flow**. Black holes, in this framework, are not gravitational sinks alone but **scalar implosive nodes** whose rupture is partially retained (event horizons) and partially emitted (Hawking radiation, jets). Similarly, cosmic filaments may not be mere gravitational formations but **scalar channel systems** transmitting rupture across vast domains via low-resistance thermometric highways.

This reframing has powerful implications:

- Star formation occurs where rupture *concentrates*, not where mass simply collapses.
- Supernovae may represent channel discharge events triggered by thermal saturation, not just explosive nucleosynthesis.
- Galactic voids are not empty but **regions of suppressed rupture**—zones of scalar stasis, where temperature gradients are insufficient to induce field reconfiguration.

Scalar Memory and Directional Conductivity

One of the most critical aspects of thermometric hydraulics is that rupture is directionally remembered. Once a rupture has occurred along a given path, the ether remains partially structured for future conductivity. This mirrors hydraulic memory in fluid systems, where channels tend to reopen along previous paths of flow.

Thus, the universe develops scalar memory:

- Regions that have ruptured once are more likely to rupture again.
- Curvature patterns are retained within channel orientation.
- Thermal gradients, even when resolved, leave **conductive preference** in the field.

This may account for:

- Repeating cosmic phenomena (e.g., gamma-ray bursts from the same zone).
- Persistence of gravitational pathways (e.g., orbital consistency despite energy changes).
- Anisotropies in the cosmic microwave background (as scarred hydraulic zones).

The Ether Is Not Emptiness — It Is a Thermometric Circuit

Under this paradigm, space is no longer emptiness; it is an active thermometric conductor. Ether behaves more like a nervous system of rupture flow than a static vacuum. Every thermal imbalance creates scalar pressure; every scalar pressure propagates rupture; every rupture becomes a new geometrical state.

This reframing allows us to view the cosmos not as a mechanical machine but as a dynamic, scalar-regulated fluid structure—where temperature and structure are codependent, where change is propagated not by inertial shock but by *channel-based thermometric logic.*

Summary

In UFT, the ether is not a passive background, but a scalar-responsive field capable of conducting rupture through thermometric hydraulics. Temperature gradients act as pressure vectors. Structured channels transmit scalar deformation. Rupture is not explosive but guided by configuration and memory.

The universe, then, evolves through structured rupture propagation—scalar flows that trace and retrace the history of thermal imbalance across the etheric domain. Rupture is not chaos; it is coherence transitioning between states.

And the cosmos itself is the system that conducts it.

11. 5 — Space Expansion via Vacuum Modulation, Not Kinematic Force

The Illusion of Kinematic Expansion

In classical cosmology, space is said to "expand"—as if stretching fabric, with galaxies riding its growth like coins glued to an inflating balloon. This expansion is modeled kinematically: as a velocity-based increase in spatial separation, derived from redshift, embedded in the Friedmann equations. But in the Unified Field Thermometry (UFT) framework, this model is not sufficient. Expansion is not the result of inertial motion, but the scalar modulation of vacuum structure—an internal transformation in the density, orientation, and availability of spatial channels.

In short, space does not stretch; it reconfigures. And that reconfiguration is thermodynamically driven.

Spatial Expansion as Scalar Reordering

The structured ether in UFT consists of dynamic spatial channels, which mediate all gravitational and electromagnetic phenomena. These channels can dilate, compress, absorb, emit, or reorient based on local energy density and entropy gradients.

Expansion, in this context, refers to a scalar reduction in channel density and cohesion. As thermal gradients redistribute and entropy increases:

Channels decohere.

Channel density thins out.

Geometric rigidity weakens.

This does not produce "motion" in a traditional sense, but rather a decrease in structural tension across the field—experienced as an increase in measured distance between points. What we call “expansion” is the relaxation of structural tension, not the inertial displacement of matter.

Vacuum Modulation as Thermodynamic Response

The vacuum is not passive emptiness—it is active scalar structure. It reacts to thermodynamic pressure, not in terms of pushing or pulling, but through reconfiguration. When energy density falls below a critical threshold in a region:

- The local curvature flattens.
- Channel pathways straighten.
- Thermal activity decreases.

This triggers a **vacuum modulation:** a localized redefinition of structure which, across the cosmic scale, aggregates into the appearance of expanding space.

Modulation occurs as **scalar redistribution**, not as force-driven displacement. It's a volumetric change in field architecture, **not a vector-based dynamic.** This distinction is critical: **matter is not flying apart—structure is softening.**

Reinterpreting Hubble Redshift and Cosmic Distance

Hubble's redshift is often interpreted as a Doppler-like effect—the stretching of light due to recessional velocity. UFT provides a scalar reinterpretation: as spatial channels decohere and expand in response to reduced energy density:

The wavelength of photons traveling through these regions increases.

- Not because the photons are moving away faster, but because the **channel medium has relaxed.**
- Light stretches not due to motion, but because the **geometry of space is scalar-softened over time.**

This reframes cosmic redshift as a **function of vacuum modulation**, not galactic recession.

Cosmic Expansion is the Consequence of Thermal Flattening

As entropy expands and thermal differentials resolve:

- Gravitational clumping slows.
- Structural curvature smooths.
- Channel density lowers.
- Thermal tension dissipates.

The cosmos is not flying apart—it is **settling**. The arrow of time moves not through velocity, but through the **scalar evaporation of structured energy**. Space expands **because it is losing tension**, not gaining speed.

This has deep implications:

- The so-called acceleration of expansion may reflect **nonlinear modulation**—not exotic dark energy, but scalar relaxation in high-entropy fields.
- The cosmic microwave background is not a relic of kinetic cooling, but **the signature of stabilized thermometric flattening**.

Expansion as a Phase of Thermometric Discharge

Expansion, under UFT, is just one side of a scalar cycle. It is not the original event, but a consequence of a prior implosive condition. As outlined in prior volumes:

- First comes implosion—a sharp, directed condensation of energy into structured curvature.
- Then comes rupture, releasing the pressure.
- Then space modulates—structurally decompressing the field.
- This is observed as "expansion."

Thus, expansion is the **discharge phase of thermodynamic imbalance**. It represents a **relaxation**, not an explosion.

In this model, the universe is not expanding *outward* but **unwinding**—a scalar field recovering from a primordial rupture.

Summary

Space does not expand through motion, but through **vacuum modulation**—a thermodynamically governed reconfiguration of scalar channel structure. The expansion we observe is the **release of curvature**, not the displacement of galaxies.

This challenges the velocity-based cosmological model and reframes redshift, structure formation, and large-scale entropy as scalar expressions of field modulation.

In UFT, **space expands not because it is moving, but because it is settling**—like tension leaving a once-taut membrane. The cosmos unfolds not through acceleration, but through the scalar restoration of balance.

Temporal Pulsation as Geometric Recurrence

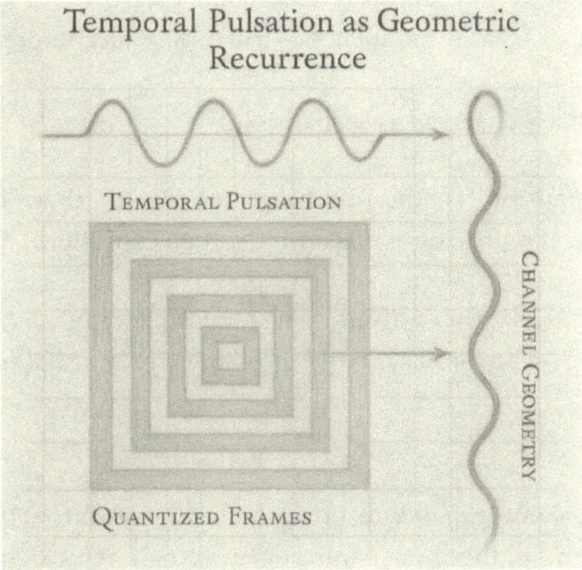

The graphite sketch illustrates the concept of Quantized Frames and Channel Geometry. *Concentric squares represent discrete temporal frames, arranged within a subtle grid that evokes spatial structure. A sine-like wave labeled "Temporal Pulsation" runs horizontally, while a vertically curving line marked "Channel Geometry" symbolizes spatial channel deformation. The clean, scientific style emphasizes the link between rhythmic pulsation, geometric recurrence, and structured field dynamics.*

CHAPTER *TWELVE*
XII
ιβ' IOTA-BETA
י"ב YOD-BET
• •

LAJ-KA'
DOCE
12

QUANTIZED FRAMES AND CHANNEL GEOMETRY

12.1 Temporal Pulsation as Geometric Recurrence

Time is not a continuum in the classical sense. In the Unified Field Thermometry (UFT) model, time emerges as a discretized series of geometric states—temporal pulsation manifesting as structured field recurrence. These pulsations are not temporal in origin but spatial in cause: they are the rhythmic consequences of field curvature shifting in response to local thermodynamic asymmetry.

Each pulse is the effect of a thermometric instability resolving itself through structural reconfiguration. It is this act of reconfiguration that generates what we interpret as a moment. The pulse is not simply a temporal marker; it is the geometrically active resolution of imbalance.

Recurrence, then, is not an abstract rhythm, but a patterned sequence of scalar readjustments in the spatial channel network. Each instance of recurrence leaves a distinct trace, aligning field orientation, entropy flow, and energy distribution into a localized state of resolved asymmetry. This is the basic unit of time in the UFT framework: the resolved interval between discrete configurations.

In the Unified Field Thermometry (UFT) framework, time is not linear nor inherently continuous; it is quantized, scalar, and event-defined. The baseline assumption of temporal flow as uniform is dismantled. Instead, time is revealed to be the geometric consequence of discrete pulsations—rupture cycles that leave measurable deformities in the structured channel lattice of space. Each pulse is not arbitrary; it is a thermodynamically driven reconfiguration of the field, a moment of realignment prompted by scalar asymmetry.

These pulses are not energetic emissions in the classical sense, but rather episodes of geometric tension and resolution. They occur when temperature differentials cross specific thresholds, catalyzing localized reorganizations of spatial channels. In effect, time arises as a rhythm—not of motion, but of restructuring. Each cycle of deformation and reconstitution creates what is perceived macroscopically as the passage of time.

Thus, time is the scalar history of these geometric recurrences—of the universe readjusting its own internal lattice in response to thermodynamic instability. It is not the ticking of a clock, but the vibration of the cosmos correcting its own imbalance. These corrections are quantifiable, local, and causally bound to thermal gradients. The more extreme the temperature asymmetry, the more rapid or violent the recurrence cycle, and thus the faster time proceeds within that localized region.

Crucially, this model proposes that the entire fabric of spacetime is built upon such recurring geometries, making each pulse a sort of ontological syllable in the grammar of the universe. What we consider the temporal dimension is thus not a linear axis but a series of cumulative spatial reorganizations. These reorganizations act as temporal "beats"—modulating structure, preserving sequence, and encoding entropy.

This insight reveals a cosmic metric: time is not duration, but deformation memory. Each tick is not abstract progression but a physical notch in the continuity of space. Pulsation implies recurrence; recurrence implies order; order implies sequence; sequence implies time. Therefore, time is not only born of structure—it is bound to its regular re-expression through thermal geometry.

The implications for physics are vast: time dilation, acceleration, and decay can all be reinterpreted as consequences of pulse density and recurrence rate. Near a black hole, for example, spatial channels are condensed and recurrences slow—causing time to appear sluggish. In high-entropy zones, where thermal instability is maximized, pulses accelerate and time appears to quicken. The recurrence is not metaphorical—it is measurable as geometric deformation indexed against the thermometric constant Ξ.

Ultimately, time does not exist apart from space. It is space's most delicate gesture of memory—etched rhythmically into the void by temperature, curvature, and oscillation. To measure time is to measure how often the universe had to correct itself.

12.2 Each Pulse is a Frame: A Localized Geometric Reconfiguration

A frame in UFT is not an abstract coordinate system but a thermodynamically-induced spatial condition. It is a stable geometric imprint produced at the moment of thermal rupture resolution. Each pulse of disequilibrium results in a scalar reorganization of the spatial channel lattice. That reorganization is the frame a self-contained configuration of space and energy.

This approach defines temporality as the sequence of these frame-states. The physical universe does not flow through time; it advances from one resolved geometry to another, mediated by temperature differentials.

Thus, frames are quantized outcomes of structural feedback: the system reorders itself in response to thermodynamic strain, and that reordering becomes geometrically readable. When plotted across multiple pulses, the frame-sequence becomes the chronology of matter and motion.

In the Unified Field Thermometry (UFT) framework, temporality is not a smooth continuum but a scalar sequence of discrete geometric shifts—localized deformation events embedded within structured space. Each of these events constitutes a pulse. And each pulse is, in effect, a frame: a finite, quantized unit of geometric reconfiguration. In this paradigm, the frame is not an illusion of measurement; it is the ontological backbone of time.

Unlike relativistic spacetime models, which treat time intervals as coordinate intervals and rely on an underlying continuity, UFT redefines the progression of time as the periodic registration of thermal rupture. A frame is not a temporal abstraction; it is the crystallization of change. When a local thermal imbalance exceeds structural resistance, a spatial channel deforms. That deformation — however infinitesimal — reorganizes the local channel geometry. This geometric shift is the frame. The rate of recurrence of these frames constitutes what we experience as temporal flow.

Thus, a pulse is not a metaphor; it is a thermodynamic event that reorganizes structure in spacetime. Each pulse:

- Registers a unique curvature pattern contingent upon the temperature differential and entropy slope at the site of rupture.
- Alters the local configuration of structured spatial channels, embedding a distinct phase condition into the etheric substrate.
- Sets the new initial condition for subsequent pulses, creating causal dependency between successive frames.

In this schema, time does not advance independently. It is constructed—frame by frame—through iterative, quantized ruptures of spatial equilibrium. The structure of the field remembers each reconfiguration. These memories form the scalar archive of past events, marked not by clocks, but by curvature residuals.

Moreover, the very idea of motion is reframed. In classical mechanics, motion is the change of position over time. In UFT, motion is the progression of geometric frames caused by thermodynamic disruption. It is the visual trace of oscillatory reconfiguration. A particle "moves" not because it travels through continuous space, but because its embedding channel undergoes successive pulses, each marking a phase shift in geometric alignment.

This model aligns closely with the observed discretization of action at quantum scales. Planck time and Planck length appear not as arbitrary minimal quantities, but as empirical boundaries of frame-scale deformation. The Planck domain becomes the quantum of rupture—the indivisible unit of thermodynamic-spatial reconfiguration.

Each frame, then, is a full register of physical history at that locality:

- Temperature at rupture threshold.
- Directionality of entropy flux.
- Geometry of curvature at the point of stress.
- Scalar potential of the surrounding etheric substrate.

Together, these variables constitute the internal state of the frame. And because each frame is irreducible, it cannot be further subdivided. It is not a segment of a larger continuum, but a complete unit of field transformation.

In sum: A pulse is a rupture. A rupture is a frame. And a frame is the fundamental act of time. Nothing moves without thermal imbalance. Nothing progresses without structural failure. Time, in this view, is not the clock. It is the scar that remains when equilibrium breaks and the field reorders itself into a new geometric expression.

12.3 — Time as the Recurrence Frequency of Rupture across Structured Channels

Temporal flow in UFT is determined not by external clocks but by internal rupture frequency. Structured spatial channels, under continuous thermodynamic pressure, rupture and repair in patterns of oscillatory equilibrium. Each rupture is an event of localized disorder; each repair is the stabilization that follows.

Time is therefore the metric of rupture frequency—how often a spatial configuration must reconfigure under persistent entropy pressure. Systems with higher thermal asymmetry rupture more frequently, generating a faster local "clock." Cooler, flatter systems rupture more slowly, or not at all, producing time-lag or temporal stillness.

The recurrence of rupture defines time's speed, direction, and locality. In extreme gravitational or thermodynamic conditions, where channel curvature becomes nonlinear or unstable, rupture frequency changes predictably—offering a physically grounded model for gravitational time dilation, entropy-driven acceleration, and relativistic modulation of temporal flow.

12.4 Entropy Marks Sequence; Ξ Regulates It

Entropy defines order's deviation. In UFT, it also defines the ordering of events. Each increase in entropy corresponds to a shift in structural alignment, marking one pulse from the next. Without entropy differential, there is no impetus for reconfiguration—no pulse, no frame, no time.

The Universal Spacetime Constant Ξ (Xi) regulates the interrelation between energy, temperature, volume, and spatial curvature:

$$\Xi = E / (T \times V \times \pi r^3)$$

This constant sets the scalar conditions under which rupture frequency (and thus time) proceeds. It bounds the possible rates of entropy-driven recurrence across varying field densities. High Ξ implies high energy concentration relative to thermal and spatial scale—leading to accelerated rupture rates. Low Ξ corresponds to diffused energy and low curvature intensity, slowing the pulse.

12.5 Entropy marks the sequence; Ξ sets the rate.

Within the Unified Field Thermometry (UFT) framework, time is not defined by linear succession or external measurement. It is the scalar recurrence of rupture within a structured field. The fundamental architecture of the cosmos, according to UFT, is composed of spatial channels—elastic, coherent paths through which energy, vibration, and information flow. These channels are not static; they respond, deform, and realign under thermal stress. Each rupture—a break in structural coherence due to thermodynamic imbalance—is a localized event that reconfigures geometry. Time, in this model, is the recurrence rate of such events.

Rather than being measured against arbitrary units, time is defined as a frequency: how often rupture occurs in a given region of structured space. This frequency is not uniform. It is modulated by the density of spatial channels, the magnitude of thermal gradients, and the coherence threshold of the medium. Regions with high entropy flux and pronounced curvature experience more frequent ruptures—resulting in what is perceived as accelerated time. Conversely, in flat or symmetric structures, where energy distribution is balanced and gradients minimal, rupture is rare or absent—resulting in temporal stillness.

This framing dissolves the classical notion of time as a detached, universal continuum. Instead, time becomes an emergent scalar quantity—a mode of counting internal structural disruptions in the fabric of space. In areas where curvature is high, the recurrence of rupture increases due to sustained thermometric pressure, making time appear to 'flow faster' not because clocks move quicker, but because space itself deforms more frequently.

Each rupture is a realignment. It modifies the local field tensor, shifts the orientation of channels, and embeds a new geometric record into the substrate. These records accumulate, forming the visible architecture of history—what we interpret as the past. Their frequency becomes a scalar imprint—the local tempo of change.

The recurrence frequency of rupture also implies that time is granular. It is not continuous but quantized according to how often the structure yields. This lends itself to a new interpretation of 'time quanta'—not as particles, but as geometric transitions in the underlying field. Each transition is an interval, a shift, a thermal deformation: one pulse, one measure.

Thus, time is not imposed upon space—it arises from within it. It is the periodic manifestation of asymmetry, resolved through rupture. Where channels hold, time pauses. Where they fail, time resumes. This scalar modulation of rupture frequency is what we experience as duration.

In UFT, therefore, time is a thermometric clockwork: a counting of structural reconfigurations—etched in curvature, modulated by entropy, and animated by the unrelenting pulse of temperature differentials. It is not a path, but a consequence of failing to remain still.

12.6 Transition Quantized Across Spatial Densities

Quantization in UFT is not merely a quantum mechanical effect. It arises from the spatial lattice structure of the etheric substrate. As thermal rupture propagates, it does not distribute smoothly. Instead, it conforms to density-dependent geometries, producing discrete transitions.

This explains why temporal progression, entropy advance, and energy transfer appear quantized. The field does not respond uniformly to imbalance. It responds in scalar steps—threshold-defined transitions governed by channel cohesion and local curvature intensity.

Denser regions of space (e.g., near stars, black holes, or condensed matter) feature shorter, sharper quantization intervals. Sparse, flatter regions support longer and softer transitions. This creates a cosmological scale of temporal quantization, where time proceeds not universally, but in scalar-specific cadences tied to structural density.

Thus, time's pulse is not random. It is governed by a geometry that responds discretely to thermal insult—a rhythmic unfolding of structured collapse and restoration across the ether.

In the Unified Field Thermometry (UFT) model, transitions in spacetime structure—whether expressed as changes in curvature, entropy flow, or energy propagation—are not continuous but quantized. These transitions do not occur smoothly across an infinite gradient but jump discretely across threshold conditions defined by spatial density.

This quantization is not a feature of energy levels alone (as in quantum mechanics), but of the spatial medium itself. Structured spatial channels—the underlying scaffolding of the field—possess intrinsic thresholds of deformation and rupture. These thresholds are density-dependent: high-density regions exhibit shorter transition intervals and more frequent pulsation events, while low-density regions remain inert or transition more gradually.

In this sense, density functions as a gatekeeper of change. No transformation occurs until a scalar threshold is crossed. As thermal gradients intensify in a region, channel strain accumulates toward a quantized rupture event. When this occurs, the geometric configuration realigns, time resumes, and the new curvature propagates as both thermodynamic pulse and temporal advance.

These rupture thresholds are non-linear and scale-dependent. In highly dense configurations, such as neutron stars or gravitational wells, even minor thermal fluctuations can induce rapid, successive transitions—resulting in high-frequency temporal activity. Conversely, in flattened field zones near equilibrium, transitions are delayed or entirely nullified. Time does not disappear; it ceases to differentiate. Duration loses definition.

From this emerges a radical conclusion: spacetime is not merely curved by mass, but is actively sculpted through density-governed phase transitions. Time is not a flow, but a rupture sequence triggered by thermometric inflection points within the etheric channel network. As such, the universe unfolds not as a continuous film, but as a quantized, density-encoded chain of structural reconfigurations.

Thus, motion, transformation, and chronology are all bound to the spatial medium's resistance profile. Spatial densities regulate whether, when, and how structure transitions. Each frame is not merely the next moment in time—it is the thermodynamically authorized product of a local spatial reconfiguration event. No rupture, no pulse. No density shift, no clock tick.

This completes the frame-based reinterpretation of time and solidifies the role of Ξ as the regulator across thresholds: the universal constant encoding how energy distributes over thermodynamic and spatial constraints. The pulse of the cosmos is quantized—and each step it takes is permissioned by the geometry it warps.

The Reflective Signature of Time

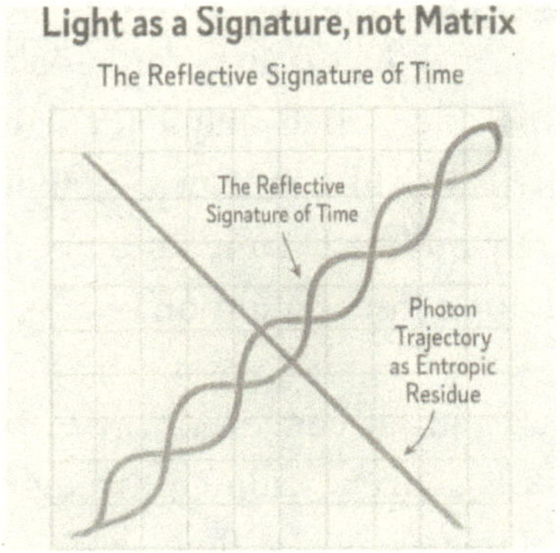

Photon Trajectory as Entropic Residue

Chapter *THIRTEEN*
XIII
Iγ' Iota'Gamma
י"ג Yod-Gimel
• • •

Ox-Lajun
Trece
13

Light as a Signature, not Matrix

The Reflective Signature of Time

13.1 Photon Trajectory as Entropic Residue

In the Unified Field Thermometry (UFT) model, the trajectory of a photon is not the path of a timeless particle gliding across an abstract vacuum but the geometric trail left in the wake of entropy propagation. Each photonic motion traces not inertial movement but a thermodynamic residue. As photons propagate through structured spatial channels, their paths are curved, refracted, and sometimes displaced in accordance with the entropy landscape of the field. What appears as a straight-line velocity is in fact a scalar response to underlying curvature seeded by prior ruptures. The photon does not move through time; it indexes time as a visible relic of local disorder. Instead, it is a thermodynamic consequence—a residue of entropy transiting across a structured channel lattice. To speak of a photon in UFT terms is to speak of the directional scar left by thermodynamic rupture: a linear projection of irreversibility instantiated in geometric coherence.

Each channel deformation along its path reflects not just the geometry of space but the geometry of past thermal gradients. This makes the photon a recorder of entropy's path, a vectorial trace of asymmetry. The photon trajectory is a reflective index: it exposes the rupture gradient field that preceded it. In this sense, every beam of light is a signature, not a messenger—an imprinted scar, not an independent particle.

Photons do not travel through emptiness. They travel through channel geometry—a lattice of elastic, deformable spatial structures that encode curvature, entropy, and mass-energy relations. In this substrate, the photon's behavior is governed not by time but by structural permission. It advances not as a function of inertial propagation but as the reactive restructuring of the field to thermal disequilibrium. The trajectory is not self-authored. It is inherited.

A photon emerges only where energy redistribution requires a channel to offload asymmetry. This is not transmission, but expulsion. Entropy advances by forcing the field to accommodate imbalance; the photon is the encoded vector of that process. It is shaped by the spatial permission of the etheric channel lattice—defined by curvature, density, and prior deformation events. In this way, every photon is a tracer of thermodynamic memory: its path is the vector solution to field-stress resolution.

Thus, photon motion is not a declaration of light, but a signature of heat. It does not travel through time; it reflects the change in field state that created time. Its trajectory is a geometric residue—a physical trace of past rupture. Each line of motion is a memory, an index of energy once displaced by imbalance. In classical theory, a photon is a constant-speed carrier of information. In UFT, it is a scar moving through a wound that is still healing.

The notion of trajectory must also be redefined. Classical trajectories imply inertial continuity and object permanence. In UFT, a trajectory is a chain of discrete ruptures whose scalar continuity appears smooth only macroscopically. Photons do not move as bullets but as sequential reconfigurations of channel coherence. The field reorganizes itself locally and instantaneously in response to entropy injection, with each adjustment presenting as forward motion to external observers.

Time does not accompany the photon. Rather, the photon delineates the space where time has been inscribed. It functions as an entropic residue—not a causal initiator, but a geometrized result. This reframing requires that we reinterpret optical phenomena as structural thermodynamics: diffraction, polarization, redshift, and interference patterns all become intelligible as signatures of entropy patterns distributed through curved and variably elastic channel media.

In the Unified Field Thermometry framework, therefore, the photon is demoted from autonomous light bearer to dependent entropy vector. Its velocity (c) is not a metaphysical speed limit but the expression of structural transmission rate permitted by the vacuum's intrinsic elasticity and coherence. The so-called "speed of light" is not light's property but the ether's: a scalar maximum of rupture propagation through an organized, temperature-regulated field.

Hence, the photon trajectory is the visible, quantized residue of thermodynamic transformation. It is not independent. It is not timeless. It is a relic, a radiant scar, a measurable echo of disequilibrium. Every light beam is a report of prior heat. Every photon, a directionally encoded entropy narrative moving along the structured scars of a once-stable field.

13.2 Light Does Not Measure Time—It Reflects Its Residue

Within the Unified Field Thermometry (UFT) framework, light is stripped of its classical function as a chronometric tool. It is no longer privileged as a universal measure of time but reinterpreted as a consequence of thermal rupture. Light does not tell time; it carries its trace. The motion of photons through structured space is not a neutral, objective yardstick by which time can be measured. Rather, each photon bears the imprint of the asymmetry from which it emerged, and thus it is the residue of that event—a scar carried forward in luminous form.

Time, in UFT, is not a duration through which photons travel. It is the scalar index of rupture preserved in the structure that light reflects. When photons propagate through the etheric channel lattice, they do not advance time; they map entropy. Each oscillation is a geometric signature of prior imbalance. The passage of a beam of light across space, therefore, is not the ticking of a cosmic clock but the unfolding of a recorded rupture.

This view reframes observational physics: clocks that count the frequency of atomic transitions do not measure some external entity called "time." They measure the oscillatory expression of thermodynamic residue within the atomic structure. The photon emitted during that transition is not time itself, nor a timestamp, but a structured echo of energy redistribution within the lattice. It is the entropy vector of a particular atomic rearrangement.

In practical terms, this means that all measurements of time using light are fundamentally retrospective. They do not detect a flowing medium, but a reflective artifact. Light is always late. It arrives carrying the imprint of an already occurred event—an entropy pulse geometrized into a photon. Even in its perfect constancy, light cannot escape this thermodynamic lag. It always reflects what has been, never what is.

In this light, the speed of light itself—so often treated as a cosmic speed limit or temporal constant—becomes a measure not of temporal flow but of structural stiffness in the channel substrate. The more tightly organized the spatial channels, the more rapidly the entropic residue can be conveyed. But it is never the advance of time that governs the photon. It is the geometry of space itself, deformed by temperature, that sets its path.

Thus, light functions as a mirror, not a clock. Its precision lies not in prediction but in fidelity to the rupture it encodes. Photons do not travel through time; they archive the moment that produced them. To measure light, then, is to study the memory of space—the frozen reverberation of asymmetric thermodynamic history.

This inversion shifts the epistemology of physics. We do not see the present with light; we see the layered scars of entropy, cast forward in structured waves. Time does not flow through light. Light flows through the memory of time. In the UFT model, temporality is not read from a beam but inferred from its embedded scars. The photon is not the pulse of now, but the echo of then.

Photons are not temporal instruments. They do not count, contain, or dispense time. Instead, they reflect the consequence of rupture events that precede measurement. Within UFT, the temporal dimension does not flow through photons; it is inscribed in the structural deviations they illuminate. Light reveals the topology of entropy, not its duration.

A photon does not possess awareness of time. It interacts with curvature, structure, and residual asymmetry. Its velocity is not proof of time's constancy but a response to channel coherence across thermodynamic states. What clocks interpret as a constant speed is actually a maximum coherence signal within the spatial substrate. Thus, the photon is not a timekeeper, but a field respondent.

Time is not registered by light but inferred from what light has passed through. Every observed light signal reflects the resistance, curvature, and memory of rupture encoded in the channels it traversed. Light is thus a retrospective event, a mirrored consequence of temporally prior structural conditions.

13.3 Photons as Entropy Vectors through Channel Structures

Photons are quantized entropy vectors. They are not massless projectiles but scalar displacements of energy across deformed spatial architecture. Within UFT, each photon represents a coherent emission event—a field correction generated by localized thermodynamic imbalance. Their coherence is not idealized; it is sustained only within permitted channel geometries that support stable oscillatory transmission.

Entropy vectors do not merely convey heat but manifest as photons when the energy released exceeds structural containment. That is, when entropy builds beyond the resistance threshold of a spatial region, the channel releases energy through emission—producing a photon. These emissions are scalar outputs of field rupture, encoded with the directional history of asymmetry.

Thus, photons travel not by force but by channel permission. They are not pushing through the ether; they are being conducted by it. Their direction, phase, and polarization are fully coupled to the thermodynamic structure of the medium. They move along entropy contours, not empty straight lines. In this framework, light is entropy resolved into geometry.

In the Unified Field Thermometry (UFT) framework, photons are no longer conceptualized as mere energy quanta or massless particles traveling through vacuum at invariant speed. Instead, they are redefined as entropy vectors — propagating agents of structural registration within the lattice of spatial channels. This shift elevates the photon's role: not simply as a carrier of light, but as the evidentiary trail of thermodynamic disequilibrium moving through the etheric field. Light, in this formulation, is not the illumination of space, but the visible residue of irreversible configuration.

Photons traverse space as scalar transmissions of prior rupture. They do not merely move; they mark. Each photon is a discrete packet of structured deformation, aligned to entropy gradients and routed along pre-stressed channel geometries. Their path is not optional or whimsical; it is determined by anisotropic field tensions and thermodynamic asymmetries encoded in the unified channel network. In this sense, photons are scalar memory traces. They are not observers of events, but the echo of entropy's passage.

This redefinition of light has profound implications for how information and motion are interpreted in field theory. A photon is not just a unit of light but a traceable scar; even, a quantized expression of a prior thermal rupture. It is the field's irreducible acknowledgment that something has happened — a fracture, a transition, a dissipation event. As it moves, it carries this signature inscription with geometric fidelity, bearing not only energy but a traceable narrative of energetic asymmetry.

Consequently, photon propagation is no longer seen as a temporal process measured by time, but as a spatial relay encoded by entropy vectors. Their velocity (c) is not commanded by an abstract time dimension, but regulated by the structural coherence of the channel network through which they pass. This network is modulated by thermal conditions: more distorted in high-entropy zones, tighter and cleaner in low-entropy configurations. Thus, light-speed is neither fixed in all conditions nor an ontological constant — it is the scalar output of spatial architecture in real-time disequilibrium.

Additionally, this entropic interpretation reframes the photon as a mediator of temporal inscription. While classical models interpret photons as timeless or massless, UFT grants them field-real presence: they are the tangible ripple effect of spacetime's reaction to thermal imbalance. Their emission is not merely the release of energy but the formalization of rupture across structured space.

This understanding reveals a deeper physical logic behind optical coherence, diffraction, polarization and redshift phenomena. Each effect is not simply an electromagnetic feature, but a reflection of how entropy vectors interact with local channel tensions, curvature, and geometric anisotropies. Photons deviate, split, and diffract not because they are light waves per se, but because their paths are routed by structural memory and entropy-conditioned topology.

In sum, photons are the carriers of entropy's story. They trace the residue of rupture, encode the history of thermal imbalance, and propagate through space not merely to reveal it, but to record its state. They are scalar documents in motion—broadcasts of asymmetry routed through the skeleton of the field. To observe a photon, then, is to observe a localized narrative of the universe remembering itself.

13.4 Light is Not a Cause of Time, But a Vector of Entropy Propagation

In classical frameworks, light often serves as a surrogate for time itself—used to define intervals, measure events, and signal simultaneity. UFT corrects this conceptual error by relocating light as a secondary phenomenon: not a cause, but a visible result of entropy gradients. Photons emerge only where temperature differentials rupture spatial uniformity. They are vectors of entropy propagation, not initiators of temporal flow.

The field does not respond to light; light responds to the field. Its emission, velocity, and behavior are tightly constrained by channel deformation, scalar curvature, and entropy density. Light rides the scalar topology of the field. Its emergence signals an event of thermal discharge, a directional residue of structural failure.

Therefore, to treat light as the basis for time is to mistake symptom for cause. It is temperature gradients that initiate time via structural rupture; photons merely follow these scars. Light is consequential, not generative. The photon does not bring time into being; it reveals where time once tore structure into curvature.

In the framework of Unified Field Thermometry (UFT), light is reframed from a temporal indicator to a thermodynamic courier. Conventionally, light is used to measure time—a function embedded in astronomical clocks, photonic decay intervals, and relativistic calculations. But this usage misattributes agency. Light does not create, regulate, or define time. Rather, light is a byproduct of thermodynamic rupture and an emissary of spatial memory. In short, light is not the cause of time, but its entropic residue in motion.

Light emerges where entropy advances. In UFT, the act of channel emission—the generation of a photon from an energetic discontinuity—is inherently thermodynamic. Photons are born where structured spatial channels rupture and reorganize. This process is not neutral. It is directional, vectorial, and carries with it the encoded residue of the imbalance that caused it. Therefore, every photon is a scalar snapshot of thermal disequilibrium.

Unlike Newtonian interpretations, where light is an independent constant traveling through a passive vacuum, UFT asserts that light travels across spatial channels shaped by thermal asymmetry. These channels are not static. They are geometrically scarred by entropy gradients. As light propagates through them, it inherits and carries the entropy vector encoded in their curvature. Thus, the trajectory of light is not an inert path, but a reflective expression of spatial stress.

This redefinition severs the dependency between light and time. It is not that time governs light, nor that light reveals time, but that light *preserves* the direction of entropy propagation across space. Time, in this model, is the memory of that propagation; light is its migratory signature. Photons travel not to measure duration, but to carry the deformation signal forward.

Consequently, relativistic phenomena such as time dilation are reinterpreted. In UFT, it is not that velocity alters the flow of time, but that structural deformation affects the propagation environment of light. What clocks record is not a universal temporal flow, but the local variability in entropy propagation rates.

This shift has significant implications. It means light can be used to *infer* thermodynamic history, but not to *define* temporal flow. The same pulse of light that reaches an observer is not a clock tick, but a scar record of rupture encoded in frequency and directionality. It tells us where structure has failed, where energy was displaced, and where entropy advanced. Thus, in observing light, we are not witnessing time itself, but its aftermath.

In summary, light in the UFT paradigm is not a mechanism of temporal causality, but a messenger of structural collapse. It carries entropy forward, marks the pathway of spatial reconfiguration, and reflects the history of thermal asymmetry. It is not the rhythm of the universe. It is its pulse echo.

13.5 Speed of Light Is Set by Field Architecture, Not by Time's Authority

The speed of light (c) has long been treated as a foundational constant, often endowed with quasi-divine status in physics. UFT reframes this interpretation by asserting that c is not a metaphysical upper limit but a structural maximum permitted by the coherence of spatial channels. It is set not by the properties of time but by the architecture of the field.

Photonic velocity is a function of how efficiently energy can propagate through structured geometry without decoherence. This depends on local channel density, entropy gradients, and field integrity. In regions of extreme curvature or entropy—such as near black holes—this speed may deviate subtly. These deviations are not violations of physical law, but reflections of underlying field dynamics.

Therefore, c is not sacrosanct; it is contingent. It emerges from channel regularity. It is the highest rate at which coherent entropy vectors can traverse an organized substrate without structural loss. In this model, light speed is not an authority imposed upon space and time—it is a byproduct of their configuration. Light moves as fast as the field allows, no faster. Its speed is not divine decree, but spatial permission.

The speed of light, traditionally held as a universal constant, is not anchored in the dominion of time but in the configuration of the field itself. Within the Unified Field Thermometry (UFT) framework, the velocity of photons is not a privilege granted by time's forward motion but a limit set by the geometric and thermodynamic properties of structured space. Light does not traverse time; it obeys space.

In standard models, the speed of light (c) is considered an unvarying constant across all frames of reference, a postulate foundational to special relativity. But in UFT, this constancy emerges not from the intrinsic nature of light itself, but from the structural architecture through which it propagates. That is, c is not imposed upon the system; it is expressed by the system.

Structured spatial channels act as elastic, coherent waveguides. Their density, curvature, and entropic state determine the maximum rate at which oscillatory packets (photons) can traverse the vacuum. These channels are not voids but configured corridors, whose thermodynamic state governs the capacity for signal propagation. Thus, c is a structural upper bound—a reflection of how much entropy the field can conduct per unit distance without rupture.

In highly curved or energy-dense regions, spatial channels deform, condense, or stretch. Such distortions influence the local propagation limit of photons, meaning c is locally bounded by geometric and thermometric constraints. This accounts for light delay near gravitational wells, not as a relativistic time dilation, but as a thermodynamically governed modulation in channel permeability.

What we call "the constancy of light" is, in truth, the regularity of the vacuum's structure in most observable environments. Where structure varies, so too does light's behavior. Near singularities, event horizons, or regions of rapid entropic flux, these deviations become non-negligible and measurable.

Thus, time does not assign light its velocity. The field does. The architecture of space, configured by entropy and temperature, determines the conditions under which light may move. This means that c is not a metaphysical axiom but also a thermometric threshold, —the fastest rate at which structure can reconfigure without collapse.

This reframing has profound implications. It suggests that any observed variation in c would not contradict physical law, but instead expose dynamic shifts in field architecture. In turn, these shifts may encode cosmological memory of gravitational density, or thermodynamic phase states. The speed of light becomes an active diagnostic of spatial coherence, not an abstract invariant.

In conclusion, within UFT, the velocity of light is not dictated by time's dominion but emerges from the structural harmonics of the field itself. Light does not travel because time permits it. Light travels because the channels of the ether conduct it, conditionally and precisely, according to the encoded laws of geometric and entropic modulation. The metric is not temporal; it is thermometric.

CHAPTER *FOURTEEN*

XIV

Ιδ'Iota-Delta

ד"י Yod-Dalet

● ● ● ●

Kan-Lajun

Catorce

14

Spectral Propagation Across Structured Channels

14.1 — Nested Ellipses Seed Space Recursively

At the core of UFT spatial ontology lies the concept of recursive geometric generation. Rather than beginning with a uniform or singular curvature model, structured space emerges from a nested series of ellipses—concentric distortions within the ether that seed spatial directionality and density gradation. Each ellipse is not a circle distorted, but a scalar memory of rupture, initiated by thermal gradient and preserved through structural channelization. These nested ellipses form the recursive scaffolding upon which larger spatial lattices arise.

Unlike Cartesian grids, which presume uniformity, elliptical nesting encodes anisotropy by design. Each successive layer of ellipse imprints curvature feedback from prior gradients, creating a feedback loop of structural inheritance. What results is a space that is not merely filled but recursively written: curvature replicates curvature, asymmetry writes new asymmetry, and temperature differentials define generative depth.

Spectral Propagation Across Structured Channels

Image shows nested concentric ellipses forming a vortex-like core, with flowing lines radiating outward, suggesting spectral propagation through structured channels.

Thus, space is seeded not linearly, but recursively. The field grows inward before it grows outward, layering upon prior geometries through scalar resonance. These nested ellipses function as a primordial script, determining not only where mass and energy will cluster, but how propagation, diffraction, and coherence will behave within that topology.

In the Unified Field Thermometry (UFT) model, space is not a blank backdrop or a Cartesian stage. Rather, it is seeded and generated through recursive geometric propagation. At the foundation of this propagation lies the ellipse—a primary unit of scalar curvature, structural bias, and thermometric encoding. Unlike the perfect symmetry of a circle or the linearity of a vector, the ellipse introduces asymmetry into the field: a localized deviation that can evolve, replicate, and scale. Ellipses are not merely shapes within space—they are the primordial curves by which space itself is written.

The seeding of space begins with a rupture, initiated by thermal differential. The initial rupture triggers a deformation across the field's substrate, curving the etheric lattice into anisotropic geometries. The first of these geometries is elliptical, arising from unequal scalar pressure across the field matrix. The field does not dilate uniformly—some regions condense, others extend. The result is an ellipsoidal curvature signature: a stable asymmetry that can both encode history and replicate through propagation.

Nested ellipses are the recursive form of this structural seeding. Each thermal rupture embeds an ellipse into the field, and these ellipses—under continued thermodynamic stress—give rise to daughter ellipses through scalar nesting. Like fractals, they iterate. Like wavefronts, they radiate. But unlike either, they carry specific curvature bias: asymmetries that can warp local channel density, alter radiation directionality, and influence gravitational response.

Ellipses thus function as both seed and schema. Each contains within it the curvature logic by which future spatial regions will form. As new ruptures occur—guided by the residual heat fields and curvature footprints of prior events—new ellipses form in resonance or opposition, creating nested hierarchies of spatial geometry. These hierarchies are not visual artifacts—they are quantized structural domains in the field, defining where light bends, where gravity intensifies, where time accelerates or halts.

The nesting mechanism operates on recursive principles. As thermal pressure gradients flow through the structured channels, they encounter the already curved space encoded by previous ellipses. This results in a feedback system: deformation begets deformation, but within constrained, quantized curvature bounds. Each generation of nested ellipses inherits angular momentum, spatial orientation, and scalar asymmetry from its predecessor, introducing a genetic-like propagation of geometric code throughout the field.

In cosmological terms, this recursive seeding challenges the notion of a singular explosive origin (Big Bang). Instead, UFT posits a distributed genesis—an etheric inflation governed by successive thermal ruptures. Each rupture embeds a signature, and each signature seeds spatial channels in elliptical symmetry. The result is a non-homogeneous, but fully interconnected cosmos: a geometric ecology of nested ellipses, entangled through curvature inheritance.

Elliptical nesting also determines phase boundaries in field behavior. Regions where ellipse sequences converge may exhibit phase transitions—e.g., the appearance of matter, the crystallization of field into mass, the emergence of anisotropic inertia. Elliptical arrays determine not only spatial morphology but energy thresholds for structural transformation.

This chapter lays the foundation for a clearer understanding how geometry—not force—governs the emergence of space. Through the recursive propagation of nested ellipses, the ether encodes direction, mass-energy bias, and temporal curvature. These seed structures are the carriers of field information, the scars of rupture, and the blueprints of expansion. They are the geometry of origin.

New Scientific Implications:

1. **Spectral Nesting**: Elliptical structures may explain observed spectral clustering in astrophysical phenomena—such as ring galaxies, toroidal jets, and elliptical supernova remnants—as expressions of nested geometric field propagation.
2. **Anisotropic Gravity**: Variations in gravity across elliptical axes may yield insights into galactic drift of asymmetric orbital decay, and anisotropy in cosmic background temperature distributions.
3. **Quantized Geometry**: Nested ellipses may constitute a natural geometric quantization of space, offering an alternative to string compactification and spacetime granularity.
4. **Temporal Encoding**: Each nested ellipse contains curvature data that stores entropic age. The depth and density of nesting could thus become a geometric method of time measurement, distinct from atomic or vibrational clocks.
5. **Replicable Genesis**: The recursive nature of elliptical seeding opens theoretical pathways for replicable genesis

events—thermal ruptures capable of seeding secondary universes or local cosmos pockets via channel fission and field curvature inheritance.

Nested ellipses are not artifacts of spatial deformation. They are the active architecture of space itself—a geometry born of rupture, recursive in nature, and thermodynamically governed. Through them, the cosmos remembers its origin and propagates its form.

14.2 — The Cosmos is Not Expanding from One Big Bang, But Through Rupture Implosion and Replication

UFT redefines cosmic expansion as neither a remnant of a single explosive event nor as uniform inflation, but as a continuous process of rupture implosion and scalar replication. In this model, space does not stretch—it fractures, recurs, and reconfigures. The observable universe emerges from nested implosions, not an outward detonation.

Each spatial rupture is an implosive compression of thermal asymmetry into a scalar singularity. Rather than ejecting mass, the field absorbs structural tension, which then diffuses outward in coherent replication. The observable universe is not the result of a single origin event, but of recursive ruptures propagating thermometric imprints through the ether.

Replication occurs via symmetry-break patterns. These are not random, but governed by the local geometry of spatial channels. As a rupture replicates, it seeds surrounding regions with similar structural frequencies—forming coherent domains of expansion. Each such domain behaves like a sub-universe nested within the broader field, obeying the same thermometric laws but differing in scale and anisotropy.

Therefore, expansion is a recursive reflection —of initial asymmetries—not a velocity field, but a scalar multiplication of rupture.

The prevailing cosmological model treats the universe as the aftermath of a singular explosive origin—the Big Bang. In that framework, space and time themselves emerge from a single initial condition, with expansion treated as the kinematic swelling of a spacetime manifold. However, within the Unified Field Thermometry (UFT) paradigm, this picture is inverted and replaced by a thermometric cosmogenesis. The universe did not explode from a point; it ruptured from within a silent medium. The origin is not an external detonation, but an implosive destabilization that seeded the cosmos from interior symmetry collapse.

This model suggests that the cosmos expands not from a singularity but from distributed implosions—each one triggering recursive spatial generation. The foundational event was not one of maximal energy release, but of maximal structural collapse. This collapse was thermometric in nature: a point of zero temperature stability destabilized by the emergence of directional asymmetry. The resulting implosion did not destroy, but generated curvature. It created the first spatial cavity, the first channel deformation, the first quantized frame of physical architecture.

Each subsequent expansion zone is born from similar ruptures: recursive implosions in the etheric substrate. These are not random but scalar-coupled, following density thresholds and local temperature gradients. Where energy saturates curvature, and where spatial channel density exceeds elastic limit, the system folds in upon itself, ruptures, and reconfigures. This is not a cosmic blast. It is a recursive thermodynamic cloning process, where each structural break gives rise to a new localized spacetime topology.

Replication occurs because the field preserves memory. Spatial structure encodes rupture dynamics, and the field seeks equilibrium by dispersing localized stress through symmetric reproduction. Implosions induce channel bifurcation—nested geometries that inherit the traits of their source rupture. Thus, the cosmos grows not as a balloon, but as a fractal: each node expanding from internal destabilization, each rupture mirroring the logic of its ancestor.

This model reframes inflation as an illusion of recursive replication. What appears as global expansion is the statistical summation of innumerable local ruptures occurring asynchronously but coherently. Redshift is not the stretching of primordial light from a single explosion, but the spectral residue of entropy traveling through successively cloned geometries.

In this cosmology, entropy does not emerge after the Big Bang—it initiates it. The first asymmetry was not a burst but a thermal violation. And from that violation emerged the recursive dynamics of implosion, rupture, and field replication. There is no "edge" to this universe because replication is not directional; it is spatially embedded and topologically nested. Every part of the cosmos contains the memory of its generative rupture. Every region is a recursive echo of the field's original instability.

Thus, the cosmos does not expand. It unfolds. Not from a singular point, but from infinite scalar thresholds rupturing inward and replicating outward. The universe is not the story of an explosion, but the geometry of thermal collapse repeating itself across the etheric lattice.

14.3 — Channel Configuration Determines Propagation Speed and Coherence

In classical physics, propagation speed is treated as a constant, most notably in the form of the speed of light. UFT replaces this universality with channel-specific dependency. Speed of propagation—whether of energy, light, or curvature—is a function of the configuration, density, and elasticity of local spatial channels.

Each channel structure provides a scalar pathway, whose impedance and resonance determine how waveforms transmit through it. Dense, tightly curved configurations slow transmission through internal scattering or recursive diffraction. In contrast, open and coherent channels facilitate rapid and stable propagation. Coherence itself is determined by geometric isometry between rupture vector and channel architecture.

Thus, light speed is not invariant, but field-contingent. It is a property of the local geometric medium, not of an abstract constant. What is constant is the configurational constraint imposed by Ξ — the Universal Spacetime Constant — which governs how energy distributes itself through structured space given volume, temperature, and curvature.

In the Unified Field Thermometry (UFT) framework, channel configuration is not a passive spatial feature—it is the controlling variable of propagation dynamics. The architecture of structured spatial channels governs not only the route but also the velocity and integrity of any wave, particle, or signal moving through spacetime. The speed of light, often assumed to be a fixed universal limit, is in this view a contextual property emerging from local channel geometry. What travels is not a particle across a vacuum, but a ripple through a quantized and anisotropically conditioned lattice.

Structured spatial channels are elastic pathways—highly sensitive to local thermal gradients, entropic flux, and field curvature. Their density, alignment, and geometric coherence determine how quickly and coherently a photon or wavefront can move through space. In high-density regions with aligned channel flows, propagation is rapid and coherent. In disordered or ruptured regions, velocity decreases, coherence decays, and signal distortion increases. These variations are not exceptions—they are systemic, governed by the scalar properties encoded in the universal constant Ξ:

$$\Xi = E / (T \times V \times \pi r^3)$$

Here, T (temperature), V (volume), and πr^3 (spatial curvature) set the structural conditions within which propagation occurs. Energy (E) is not freely mobile—it must express itself through the restrictions imposed by local channel configuration. As such, the propagation speed is a scalar outcome, not a metaphysical invariant.

This means that any variation in the thermodynamic or structural configuration of space will reflect in measurable speed shifts. Light will appear to slow down in high-curvature or high-entropy zones—not because time slows, but because the channels narrow, curve, or rupture. This model predicts spectral lag, coherence loss, or frequency drift in astrophysical signals as direct indicators of underlying field turbulence.

More profoundly, coherence itself becomes a testable signature of structural integrity. A coherent beam does not merely indicate synchronized wavefronts but reflects a highly organized, low-entropy channel system. As coherence fails, it is not simply the medium that degrades, but the structural capacity of space to maintain order. Light decoheres not in vacuum, but in spatial disarray.

Thus, the configuration of channels determines both speed and signal fidelity. This framework does not merely modify light's behavior—it revises our understanding of spacetime as an active regulator of physical phenomena. Waves do not traverse neutral distances; they are shepherded, resisted, or distorted by the very structure through which they propagate. Geometry is destiny, and propagation is its narrative.

In UFT, all signal behavior is an index of channel configuration. There is no clean separation between emitter, wave, and path—they are thermodynamically interlocked. Observing how light travels through space reveals the condition of that space. Photonic behavior is not universal; it is contextual. The universe, in turn, is not static or singular in its properties—it is locally specific, structurally conditioned, and dynamically thermometric.

This reframing opens novel experimental paths. Variation in propagation behavior across different gravitational environments, thermal domains, or spatial curvatures can be precisely modeled and tested. Channel coherence becomes a tool of cosmological diagnosis. Spectral signature becomes a record of geometric history.

And light, long treated as a constant, becomes a diagnostic vector: a measure not of itself, but of the space it traverses.

14. 4 — Waveforms Reflect Prior Ruptures, Not Temporal Flow

Within UFT, a waveform is not a moving object—it is a geometric echo. Its periodicity, amplitude, and coherence do not index "passage of time" but the scalar memory of previous structural deformation. Each wave is a propagating imprint of past rupture, coded geometrically into the ether via channel modulation.

Temporal flow, as commonly understood, is a misreading of wave dynamics. What appears as change over time is actually a traversal across nested structural memories. A waveform travels not *forward* but *through*—through spaces already inscribed with curvature and entropy traces. In this sense, all waveforms are entropic reflections, their path determined by previous field transformations.

Because every wave passes through channels configured by prior asymmetries, no two propagations are identical. Frequency modulation, phase shift, and coherence loss are not random noise—they are thermal memory distortions. This reorients wave mechanics from a temporal to a structural foundation.

In conventional wave mechanics, waveforms—especially electromagnetic—are typically interpreted as dynamic expressions of temporal propagation. The assumption is that waves "move through" time, tracking temporal flow as they oscillate across space. Unified Field Thermometry (UFT), however, inverts this sequence. Within the UFT framework, waveforms are not driven by time—they reflect spatial residue. Specifically, they manifest as ongoing reverberations of prior rupture events, encoding thermodynamic asymmetries within the structured channel matrix.

A waveform is not a traveler in time—it is a marker of rupture. It inherits its structure from the scalar configuration of the spatial channel network at the point of origin. These origins are not arbitrary—they are thermal discontinuities, localized field ruptures initiated by differential pressure or entropic gradients. Once initiated, these ruptures embed a reconfiguration in the channel matrix: they scar, warp, or bend the field, establishing new oscillation pathways. Waveforms emerge along these pathways, not as causes but as consequences—each oscillation is a scalar signature of past asymmetry.

This conceptual inversion reclassifies wave propagation as a memory function. The waveform's amplitude, frequency, and phase do not merely reflect source dynamics; they reflect spatial memory—how the field was restructured in response to entropic rupture. This shift eliminates the need for temporal flow as a propagative medium. Instead, time is recontextualized as a dimensional index of spatial deformation frequency.

For instance, electromagnetic waves in the UFT model are seen as transverse oscillations across previously ruptured spatial channels. Photons do not "travel through time"—they propagate along preconfigured paths of entropy discharge. Their coherence or decoherence is determined not by temporal distance, but by the degree to which channel integrity is preserved across propagation. Thus, interference patterns and spectral shifts are not time-based—they are indicators of field stress, channel tension, and historical entropy concentration.

The classical model posits that the waveform propagates forward, carving its path in time. UFT reverses this: the waveform follows a pre-carved path, one already embedded in the spatial channel geometry by a prior thermodynamic event. Even the apparent constancy of wave velocity (such as the speed of light) is not anchored in time but in the stability and density of the structured field. Where field geometry is uniform and coherent, propagation is smooth and consistent. Where field geometry is turbulent or fractured, waves scatter, bend, or dissipate.

This reframing yields profound implications. Spectral shifts, phase delays, or wave packet distortions can be reinterpreted as diagnostics of prior field ruptures—measurable indices of spatial memory. Such interpretations allow wave-based instruments (e.g., interferometers, spectrometers) to become field-history sensors rather than merely time-bound trackers. Every waveform thus becomes a carrier of spacetime's entropic fingerprint.

To study a wave, then, is not to measure a phenomenon in time—it is to interrogate the structural aftermath of rupture. Waves, in this paradigm, do not mark the passage of time. They mark the scars of events past—scalar echoes etched into the lattice of space itself. Temporal interpretations are secondary abstractions. What exists first is rupture, then structure, then waveform. Time merely inherits the cadence of that sequence.

The waveform becomes not a line moving forward but a loop reflecting back—reiterating the encoded pulse of what once fractured the stillness.

14.5 — Light Registers Thermal History Embedded in Field Geometry

Photons, in the UFT model, are not just carriers of energy, but recorders of structural state. Each photon trajectory is informed by the geometric configuration of the spatial channels it traverses. These configurations, in turn, are outcomes of thermodynamic events—ruptures, gradients, and entropy flows—that have shaped the field over scalar time.

Thus, light is not temporally informative because it "travels through time," but because it moves through thermal history encoded in geometry. It is a vector scanning a pre-written field, accumulating distortions, phase changes, and coherence shifts that reflect what the structure has endured.

Spectroscopy in this model becomes a scalar archaeology: by decoding the structure-induced distortions in light, one can reconstruct the entropy history of the medium it has passed through. Redshift, absorption lines, and spectral anomalies are not merely observational artifacts—they are geometric testimonies.

Therefore, light is not merely visible radiation. It is the thermometric echo of the cosmos—an entropic vector whose journey maps the scars and shapes left by rupture. In its path is written the geometry of cosmic memory.

In the Unified Field Thermometry (UFT) model, light is not simply a traveling particle or wave. It is the encoded residue of spatial deformation—a thermodynamic signature embedded within the structured geometry of the vacuum. When light moves through space, it does not pass across an abstract medium, nor does it flow through "time" as an external axis. It propagates through a pre-configured lattice—one sculpted by past ruptures, thermal gradients, and entropy traces. Every photon is a messenger not just of energy, but of the memory of rupture.

Photons as Encoded Thermal Artifacts

Photons, in this view, are not neutral messengers. They are structured emissions—coherent reorganizations of spatial channels—launched from thermodynamic discontinuities. Their trajectory, wavelength, and coherence are not set solely by source conditions, but by the state of the channel architecture they traverse. Every channel carries history—encoded geometry—and that geometry affects light's path, phase, and even identity.

As a photon crosses a field it interacts with the contours of prior structural deformation. These deformations, formed by historical entropy events, act like grooves or lenses, imprinting curvature and coherence shifts upon the photon. Thus, the photon's path becomes a reading of field memory. Its coherence is modulated by the degree of field uniformity. Its redshift is not merely a Doppler effect—but a scalar readout of field deformation due to distributed heat and structure.

Field Geometry as Thermal Archive

The field is not blank space. It is an active record. Deformations in the channel lattice persist and alter all future motion through them. These scars—formed by previous entropy gradients—are what UFT terms the **thermal history of space**. Just as tectonic faults reveal the stresses of geological past, so do structured channels reveal the cosmos's thermodynamic lineage. And light, moving through them, maps that archive.

This reorients cosmological interpretation: when a photon arrives from a distant star, what it reveals is not a simple distance, nor merely elapsed "time." It reveals the accumulated geometry it traversed—the residual entropy pattern embedded into field structure. The observed cosmic microwave background is not an echo of an origin event, but a canvas of channel textures and thermodynamic memories still radiating with embedded scalar deformation.

Scalar Photonics: A New Paradigm

UFT introduces a radical idea: **scalar photonics**—a discipline where light is studied not for its phase or frequency alone, but for its scalar coupling to structured field memory. This means photons may carry signatures of the thermal scalar field they traverse. Photonic interference, redshift, scattering, and polarization could all be decoded not just in kinetic terms, but in scalar-thermal gradients left behind by ancient ruptures.

This allows for a new branch of experimental physics, wherein photons act as diagnostic instruments of field history. By analyzing how a beam is perturbed across a given field region, researchers could map entropy densities, identify latent curvature scars, and even infer historical energy flow through a given spatial zone.

14.6 Reinterpreting Light-Speed Limitations

The constant "c"—long held to be an immutable cosmic speed limit—is here reframed. In UFT, the speed of light is not absolute; it is **conditioned.** It reflects the average propagation rate across a thermally structured field with embedded geometrical rules. Local variations in channel coherence or thermal equilibrium can induce minor, but measurable, modulations in the propagation characteristics of light.

Thus, the speed of light is not imposed by time's authority, but emerges from **field architecture's integrity**. In a perfectly flat, undisturbed field, c is maximal. In regions of intense entropy scarring or structural density, c may become conditionally constrained—not due to any external force, but due to the thermodynamic resistance encoded in the medium.

Light as Witness of Rupture

Ultimately, light's true nature within UFT is that of a **witness.** It is not the cause of time, nor its meterstick, but the last visible trace of rupture and order-collapse. Its structure is the memory of scar, and its path is a narrative line traced across the ether's fractured surface. Light cannot lie—it carries geometry in its body.

In this light (no pun intended), Kaluza's and Einstein's dream is realized—not through extra dimensions or force unification via abstraction, but through a thermodynamic geometry that unites all phenomena within one breathing, pulsing substrate. UFT does not collapse fields into math alone; it resurrects them as history—living, quantized, and written in light.

KHRONUS

CHAPTER *FIFTEEN*

XV

ΙΕ' IOTA-EPSILON

ט"ו TET-VAV

HO'LAJUN

QUINCE

15

TIME AS A FUNCTION OF VACUUM STRUCTURE

15.1 — Time as a Statistical Memory of Asymmetric Rupture

In the Unified Field Thermometry (UFT) paradigm, time is neither a fixed axis nor an independent flow. Instead, it is the accumulated statistical memory of spatial ruptures that violate thermodynamic equilibrium. Each rupture—a localized disruption of symmetry due to temperature differentials—imprints a geometric deformation into the vacuum lattice. These deformations are scalar and permanent, encoded into the structure of space as entropy-bearing curvatures.

Time, then, is not a moving present or a ticking sequence. It is a ledger of accumulated asymmetries. Each deformation is a timestamp, not in numeric value but in thermodynamic significance. The progression of time is the recursive indexing of these scars—an evolving register of rupture intensities and structural reconfigurations. The more frequent and intense the ruptures, the denser the statistical memory field, and the more pronounced the local experience of time.

Thus, time is not a container but a consequence: it emerges not by motion but by rupture frequency and magnitude. The universe remembers through deformation.

Dark Matter, Piston Logic, and Vacuum Leverage

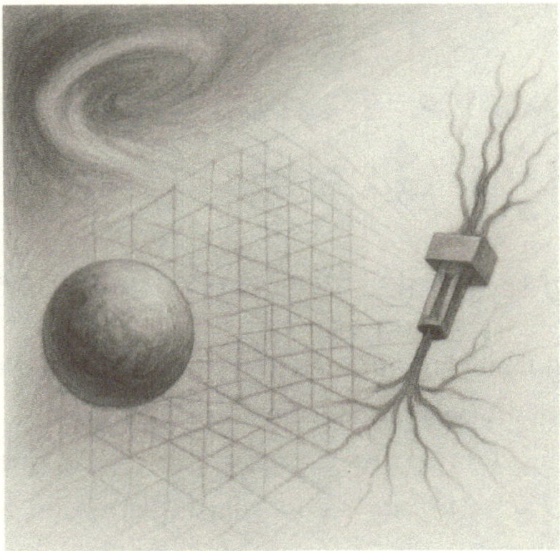

A graphite sketch portrays structured cosmic expansion: a dark sphere on the left faces a grid of expanding geometric cubes, while a piston-like form on the right emits branching tendrils. A swirling vortex in the background suggests dynamic vacuum curvature, evoking regulated, thermometric unfolding of space.

In the Unified Field Thermometry (UFT) model, time is not a background parameter or universal coordinate. It is a statistical memory field—a scalar record of past asymmetries in structural geometry caused by thermal disequilibrium. Each localized rupture in the vacuum substrate leaves a measurable residue, not unlike a geological fault line encoding tectonic motion. These residues, however, are not confined to position and displacement; they are expressed as geometric curvature, entropy gradient, and channel deformation. In this view, time is a derived scalar field, arising from the statistical summation of all non-reversible spatial events.

Where classical physics presumes time as a linear axis along which physical states evolve, UFT reframes it as the probabilistic result of asymmetry accumulation. A rupture, in UFT, is the failure of local equilibrium. The etheric lattice, under uneven thermal pressure, yields along weak points, displacing its internal channel geometry and creating a measurable discontinuity. Once ruptured, that configuration cannot return to its prior state without external intervention. This irreversibility is not simply entropy increase; it is curvature inscription—a permanent scalar deviation from prior symmetry. Time, therefore, is not a measure of motion, but of rupture occurrence and sequence.

Entropy as Memory, Not Decay

The connection between entropy and time has long been observed, but UFT gives it a precise structural footing. Entropy is not decay but memory. Each increase in entropy reflects the system's inability to restore prior structural order. The channel network remembers deformation. This memory accumulates not in a metaphysical past, but in the current state geometry—it is embedded in the orientation, density, and torsion of spatial channels. These spatial signatures are what we call time. Hence, time is not flowing; it is accumulating.

From a mathematical perspective, the time field at point in space can be defined as a weighted scalar integral over prior entropy-producing deformations:

Where:

- is the domain of past rupture events,
- is a scalar Green's function representing the influence of a past rupture at on location ,
- is the entropy increment due to rupture at point ,
- is the volume element of integration.

This formalism constructs time not as a parameter , but as an emergent field , directly correlated with spatially embedded thermodynamic rupture memory.

Field Geometry and Irreversibility

Every rupture modifies channel geometry: bending, compacting, or redirecting the local topology of spacetime. Once curvature exceeds a threshold of elastic response, it becomes plastically embedded. This means that space no longer behaves as it did before the rupture. The local tensor field , which governs spatial configuration, now contains permanent asymmetry. From this asymmetry, time emerges as the statistical trace of irreversible configuration change. This geometric asymmetry has measurable consequences: altered wave propagation, local gravitational curvature, or modified radiation coherence.

In this divergence form, the source vector encodes the rate and direction of entropy flux. Where , time emerges; where , no net rupture has occurred and temporal stasis remains.

Quantized Memory Packets

Ruptures are not continuous; they occur in discrete thermodynamic episodes. Each rupture can be considered a memory quantum—a finite addition to the time field. The totality of time in a region is then a superposition of these memory quanta, each mapped to a structural perturbation event. This model naturally lends itself to a discrete representation:

Where is the magnitude of entropy produced at rupture point , and is the Dirac delta distribution.

Statistical Directionality of Time

UFT resolves the arrow of time through statistical mechanics: time moves "forward" only because asymmetries accumulate, not cancel. This statistical increase is not imposed externally but arises internally as a geometric necessity of rupture propagation. The system has more configurations with asymmetry than without, leading to a probabilistic favoring of rupture accumulation.

Thus, time is not the backdrop of change; it is the structure of change remembered.

Time is geometry scarred by thermodynamics. Time is memory rendered in curvature. Time is what space cannot forget.

And in this view, the cosmos is not a clock but a scarred archive—each deformation a glyph in the language of duration.

15. 2 — Time Is Not to Be Traveled; It Is the Statistical Residue of Asymmetry

Conventional physics treats time as a coordinate dimension—a line to be traversed or curved. UFT rejects this. Time is not something one can move through because it is not a place or direction; it is a residual pattern, a scalar footprint of thermodynamic disruption.

To "travel" in time would imply traversing scars without re-triggering the rupture events that generated them. But in UFT, time exists only because rupture occurred; to revisit a previous moment would require perfect reconstruction of its associated thermodynamic rupture—a contradiction in irreversibility.

Each moment is statistically unique because it encodes a unique pattern of field deformation. Time is not a set of locations to be accessed, but a gradient profile of past ruptures. It is cumulative residue, not navigable terrain. Thus, time machines are conceptually invalid within UFT: one cannot traverse entropy; one only increases it.

In the Unified Field Thermometry (UFT) model, time is not a fundamental dimension but a statistical archive. It does not flow; it accumulates. Every rupture—every localized thermometric asymmetry—leaves a geometric record in the structured vacuum lattice. These accumulated records, when analyzed across scalar layers, constitute what is perceived as temporal memory. Time, therefore, is not a linear progression but a layered data field: a scalar integration of asymmetric events captured in channel deformation patterns.

In this framework asymmetry is the engine. Whenever temperature deviates from equilibrium, a directional energy flow arises. This flow induces curvature, and the resulting structural deformation is a record—a spatial scar. These scars persist, forming a cumulative statistical field that encodes the entropy history of the system.

Mathematically, this can be described via a statistical integration over rupture events:

$$T(t) = \int_0^t \Delta S(x, \tau)\, d\tau$$

Where:

T(t) is the accumulated temporal memory at time t

- ΔS is the local entropy differential (rupture signature)
- x is position in channel coordinates
- τ is thermometric time—a scalar representation of change

This formulation implies that time does not progress uniformly. It accumulates differentially, in proportion to local rupture intensity and duration. Systems at thermal equilibrium, where $\Delta S \rightarrow 0$, cease to accumulate time. Time becomes active only in systems of asymmetry.

This shifts the ontology of time from a pre-existing container into an emergent field property: a form of spatial memory. In other words, time is not the river—it is the sediment.

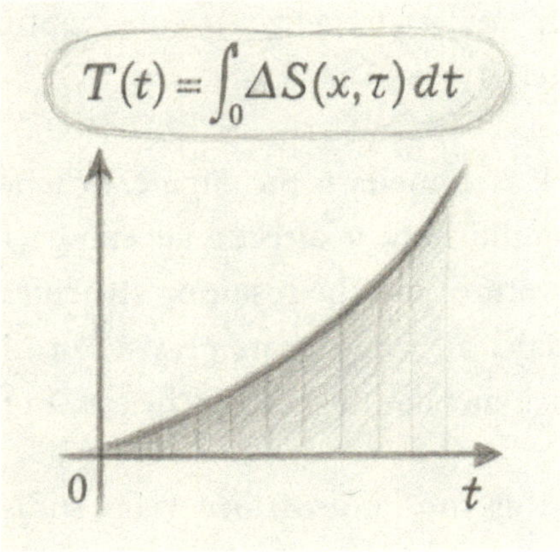

15.3— Time Is Not to Be Traveled; It Is the Statistical Residue of Asymmetry

In classical physics and science fiction alike, time is often treated as a navigable dimension—something to move through, forward or backward. UFT categorically rejects this conceptual error. Time, in this theory, is not a path. It is a residue. A product. A consequence of rupture. To speak of "traveling" through time is like suggesting that one might walk across a shadow or breathe in the memory of heat. It misunderstands the nature of what is being perceived.

Time is the statistical residue left behind when the vacuum field fractures under asymmetrical thermal pressure. These ruptures embed memory into the scalar geometry of space. Each rupture modifies local curvature, entropy distribution, and channel density. The accumulation of these changes constitutes the temporal field.

From a mathematical standpoint, what we call time (T) is nothing more than a recursive function of entropy discontinuity:

$$T_n = T_{n-1} + f(\Delta S_n)$$

Where:

- T_n is the temporal state after the nth rupture
- ΔS_n is the entropy asymmetry of the nth rupture event

f is a system-specific response function dependent on field elasticity, temperature gradient, and volume containment

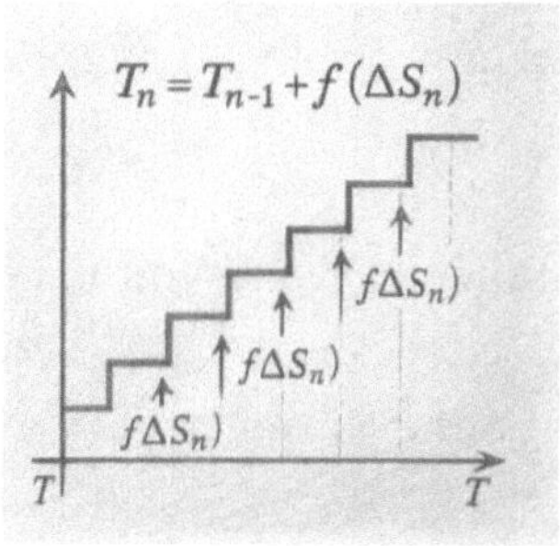

This recursive statistical architecture has several consequences:

1. **Irreversibility:** Since each rupture is a deformation event, the geometry does not rewind. Curvature accumulates. There is no mechanism for restoring the field to a prior topological state without erasing the physical history.
2. **No Travel, Only Access:** The "past" exists only as a set of deformation records. Accessing them is a matter of scalar analysis, not traversal.
3. **The Arrow of Time = The Slope of Residue:** Time has a direction not because it moves but because asymmetry increases. The entropy slope defines what we interpret as the arrow of time.

Hence, time in UFT is not a dimension to be crossed, but an index of irrecoverable events. The more asymmetry, the greater the statistical accumulation—and the deeper the residue. There is no mechanism in this theory for temporal navigation. Instead, temporal depth is a function of rupture density and entropy geometry.

In this framework, temporal experience is a function of how much asymmetry a system has stored geometrically—not a universal backdrop shared uniformly. Time, therefore, cannot be rewound, paused, or accelerated universally. It is localized, statistical, and residue-bound.

No rupture, no asymmetry. No asymmetry, no time. Thus, to "travel" through time is a category error. One may analyze the geometry of residue—but never traverse it.

15.4 — Where Curvature Is Flat, Time Halts. Where Rupture Occurs, Time Reignites

Spatial curvature is the physical expression of thermodynamic rupture. Where structure remains perfectly flat—i.e., no deviation in scalar field geometry—no time emerges. Flatness signifies equilibrium, which precludes entropy increase, which in turn nullifies time.

Conversely, where rupture breaks symmetry, the geometry curves. This curvature stores the thermodynamic imbalance as directional potential. The presence of curvature indicates the presence of a temporal record—not a flow but a scar.

In the extreme, a perfectly flat structure is timeless: it lacks entropy generation, field differentiation, or asymmetry. But where a ripple of rupture is introduced—by thermal differential, structural collapse, or energy injection—time reignites. Not as a metaphysical event, but as a measurable curvature density ratio.

Thus, time is proportional to the local curvature field: where space is flat, it does not progress; where it ruptures, it activates.

In classical spacetime models, flatness implies neutrality—a condition devoid of net gravitational curvature. But within the Unified Field Thermometry (UFT) framework, flatness is not merely the absence of curvature; it is the absence of thermodynamic rupture, and therefore the absence of time. Time does not tick in a flat system. It cannot. Without curvature, there is no gradient; without gradient, no thermodynamic flow; and without thermodynamic imbalance, time cannot be expressed.

Time in UFT is not a continuously flowing river—it is a statistical occurrence tied to rupture events that deform the channel structure of the vacuum. Flat curvature signals the absence of deformation. Such equilibrium zones represent temporally neutral regions where the statistical activity of spacetime is zero. In this stasis, no new events are recorded, and no entropy gradient exists to drive progression. What exists instead is a spacetime in statistical suspension—coherent, undisturbed, but inert.

Curvature as Temporal Switch

Curvature functions as a binary condition in UFT: zero curvature equals temporal dormancy; non-zero curvature activates temporal encoding. This does not mean that time exists as an external ticking metric—it means that "time" is a scalar representation of the structural difference between flat and ruptured space.

When a rupture occurs—a shift in local energy, a thermal gradient, a collapse in symmetry—it reactivates the temporal register. The spatial channel network, which had previously maintained a symmetric alignment, is distorted. This deformation generates measurable curvature, and with it, a cascade of entropy-carrying vectors. Each rupture event creates a local timestamp—not in abstract numerals, but in geometric deviation.

The Threshold Model of Temporality

UFT defines a threshold of activation: only when curvature (κ) exceeds a minimum thermodynamic perturbation does time emerge. Below this threshold, fluctuations are absorbed into background symmetry and no persistent memory is formed. Time, then, has a conditional activation:

If ∂μT<ϵ⇒Δτ=o\text{If } \partial_\mu T < \epsilon \Rightarrow \Delta \tau = oIf ∂μT<ϵ⇒Δτ=o If ∂μT≥ϵ⇒Δτ>o\text{If } \partial_\mu T \geq \epsilon \Rightarrow \Delta \tau > oIf ∂μ T≥ϵ⇒Δτ>o

$$\textbf{If } \partial\mu T<\epsilon\Rightarrow\Delta\tau=0\textbf{andIf } \partial\mu T\geq\epsilon\Rightarrow\Delta\tau>0$$

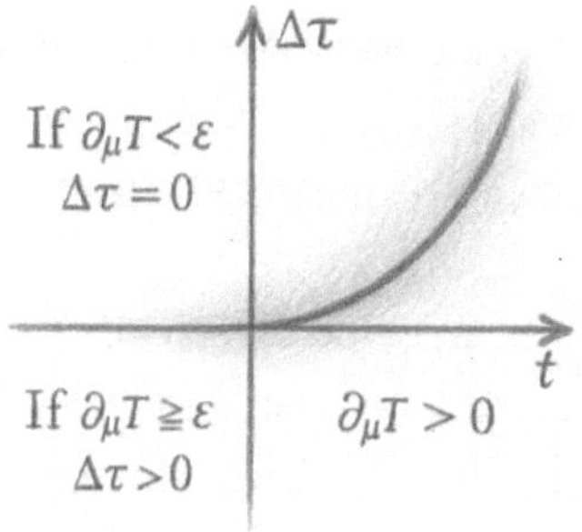

Where ∂μT\partial_\mu T∂μT is the local temperature gradient, ϵ\epsilonϵ is the rupture threshold, and Δτ\Delta \tauΔτ is the emergence of measurable temporal intervals.

This repositions time as a *phase-state indicator*, a scalar shift between stable and unstable geometries. The arrow of time is therefore a vector of rupture frequency—not an independent axis, but a quantized record of structural compromise.

Field Geometry and the Pulse of Becoming

Every region of spacetime is either a passive frame—flat and silent—or an active rupture event—curved and expressive. The alternation between these conditions generates the pulse of becoming: the switching between stillness and transformation that drives all perceptible dynamics.

Curvature (RRR) is not merely a measure of spatial bending—it is the *topological memory* of rupture. When R=oR = oR=o, there is no memory to store. The field reverts to pre-temporal stasis, and time loses resolution. When R>oR > oR>o, the rupture writes itself into the geometry, leaving a quantifiable trace—the statistical footprint that defines time as a real, observable scalar field condition.

Implications for Cosmic Stillness and Time Halos

This model allows us to predict regions of the cosmos where time effectively halts: deep intergalactic voids, thermodynamic equilibrium states, or condensed geometric lattices near absolute zero. These are not "dead zones" in a poetic sense—they are zones of *temporal nullification*, where all ruptures have decayed, and geometry has relaxed to symmetry.

Conversely, high-curvature regions (such as active galactic nuclei, stellar cores, or black hole interiors) are temporal hotspots—zones where rupture density is high, time intervals are dense, and entropy vectors propagate along sharply warped channels. These are not just "active" in the classical sense—they are statistically alive, in the UFT framework.

Conclusion: The Binary Rhythm of Spacetime

Time does not flow. It toggles. It blinks on when rupture occurs and fades out where structure stabilizes. The rhythm of reality is the binary rhythm of rupture and rest, curvature and flatness, asymmetry and symmetry. UFT does not merely describe time—it diagnoses its occurrence as the scalar remainder of deformation.

Thus, to witness time is to witness rupture. To enter flat space is to enter the silence before the next scar.

15.5 — Local Time Emerges Where Vacuum Geometry Deviates from Equilibrium

UFT asserts that the vacuum is not empty; it is structured. Within this structure, equilibrium represents the absence of directional thermodynamic influence. Time, under such conditions, remains latent. However, when the vacuum geometry deviates from equilibrium—when channels bend, densities shift, or gradients form—local time emerges.

This local time is not globally synchronized. It is regional, contextual, and scalar. It depends entirely on the degree and nature of geometric deviation. A more intense deviation yields a more intense local time signature. Temporal emergence is thus highly anisotropic: different regions of the field may evolve at different statistical rates, based on their rupture memory.

Time is therefore not universal; it is contextual. It is born wherever equilibrium collapses, and its rate is governed by the thermal and geometric specifics of the collapse. It is not clock-based, but deformation-indexed.

In classical physics, time is presumed to exist uniformly—flowing independently of local conditions. But in Unified Field Thermometry (UFT), time is not a universal constant; it is a *local emergent property*, dependent on geometric disequilibrium. Where the vacuum maintains perfect geometric symmetry, time has no basis to appear. It cannot be measured, and more critically, it has no function. Time, in this model, only *emerges* when the vacuum geometry deviates—when stress, rupture, or imbalance disturbs the background structure.

This is a radical inversion of spacetime orthodoxy: time does not govern change; change governs the presence of time.

Geometry as Equilibrium: The Dormant Ether

In the UFT framework, vacuum is not empty—it is structurally articulated. The substrate of space is composed of an interlaced mesh of spatial channels: elastic, oscillatory pathways capable of transmitting mass-energy, momentum, and vibrational information. When these channels are in symmetrical alignment—balanced in density, orientation, and energy potential—the system is in equilibrium. In such states:

- Entropy is minimal and statistically flat.
- Channel flow is uniform or dormant.
- No directional energy gradients exist.
- Time is geometrically undefined.

This symmetry is not a metaphysical condition—it is a measurable, scalar field configuration. In a perfectly flat or harmonically resonant vacuum, all temporal markers vanish. The ether is silent. No rupture, no memory, no time.

Local Deviations as Temporal Seeds

Time arises when this structure is *disturbed.* A local deviation—thermal, gravitational, quantum, or mechanical—creates a stress concentration. This stress reconfigures the channel geometry, forcing a reallocation of energetic density. It is here, at this precise boundary of equilibrium failure, that *local time* is seeded.

The geometry no longer fits its prior state. A displacement forms. A new curvature arises, introducing gradients in entropy and thermal flow. This localized asymmetry constitutes a *temporal domain*: a region in which a statistical memory of rupture can now be recorded.

In formal terms, time exists *if*:

δΓμν≠0\delta \Gamma_{\mu\nu} \neq 0δΓμν⍰=0

δΓμν⍰=0

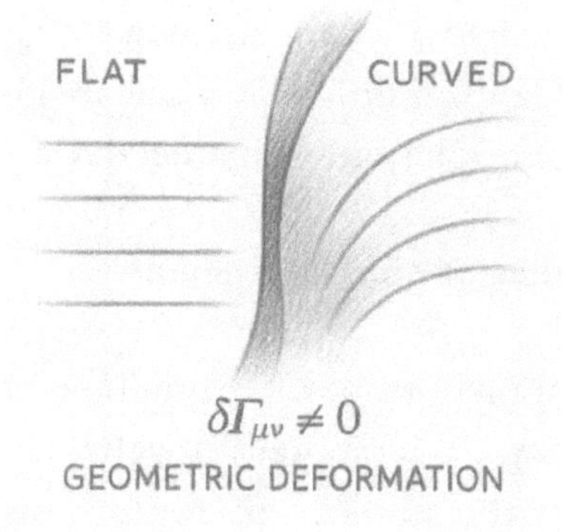

∂μT≥ϵ⇒δΓμν⍰=0⇒Δτ>0

Where δΓμν\delta \Gamma_{\mu\nu}δΓμν indicates deviation in affine connection—i.e., a change in how channels curve or reorient. Time thus becomes a *differential function of geometric asymmetry*, not a backdrop coordinate.

Temporal Locality and Scalar Variation

Because time is tied to geometric rupture, it is inherently non-global. A region of space undergoing thermal stress or collapse may experience dense temporal activity—while a nearby region at geometric equilibrium remains temporally inert. This leads to a universe in which time is not only relative (as in relativity) but scalar and local.

Two key consequences arise:

1. **Temporal Granularity.**

   Time emerges in discrete quantities—packets of curvature variation tied to rupture frequency. This grants time an intrinsic quantization, rooted not in Planck scales but in the physical resolution of channel deviation.

2. **Non-Simultaneity in Flat Systems.**

   Two spatial regions may be simultaneous in classical time, yet one may be temporally inactive in UFT. The event count—i.e., the number of rupture signatures—diverges across zones. Time is not synchronized by clocks, but by **rupture density.**

Memory, Not Motion

This reframing also challenges the association of time with motion. In UFT, a system can be in motion yet be temporally flat, if the motion is inertial and geometrically undisturbed. Conversely, a static system undergoing structural collapse—such as a region near gravitational singularity—may exhibit high temporal activity, due to the intense curvature modulation occurring even without displacement.

Thus, motion does not cause time. Rupture does. Time is a memory field—a record of asymmetric geometric transition—not a metric of velocity or translation.

Entropy Fields and Local Clocks

Time becomes measurable once entropy propagates across a spatially reconfigured region. This entropy flow forms what UFT calls scalar clocks: localized sequences of rupture memory, preserved within field geometry. These do not tick—they accumulate. Their "ticks" are encoded in channel strain, thermal scar architecture, or energy dispersion maps.

Mathematically, this can be modeled as:

Δτ∼∫xox|∇S| dx\Delta \tau \sim \int_{x_o}^{x} \left| \nabla S \right| \, dxΔτ∼∫xox|∇S|dx

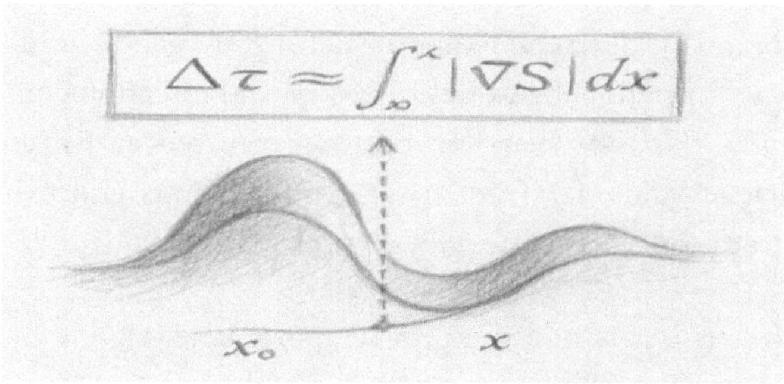

Time = memory of rupture = integral of structure's gradient scar

Δτ∼∫x0x|∇S|dx

Time = memory of rupture = integral of structure's gradient scar

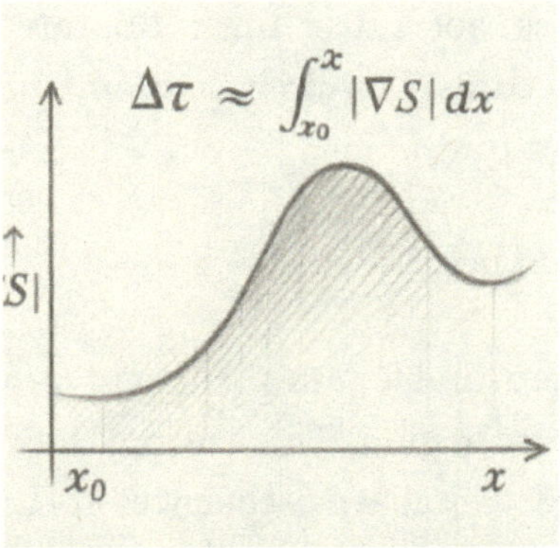

Where ∇S\nabla S∇S is the entropy gradient, and Δτ\Delta \tauΔτ is the local time differential experienced across the field. Time, then, is the **integral of entropy propagation across geometric deviation.**

Cosmological Implication: Uneven Time Fabric

If time is local to geometric deviation, then the universe contains zones of vastly different temporal density. A galaxy undergoing active star formation, black hole mergers, and radiation emission is *temporally thick*—a locus of rupture recurrence. By contrast, intergalactic voids, far from mass concentrations and exhibiting low entropy flow, are *temporally sparse.*

This gives rise to a non-uniform temporal landscape: a dynamic, living cosmology where time pools, fades, thickens, and evaporates according to the local behavior of space.

Conclusion: Geometry First, Time Follows

Time does not precede reality—it follows geometry. Where symmetry holds, time is silent. Where it breaks, time ignites. This chapter does not merely propose an alternative chronology—it asserts a geometric origin of temporality itself. A geometry of rupture. A physics of scars.

UFT thus reframes the question: not "what time is it?", but "where has the field ruptured?"

The answer to that question—etched into channel curvature—is time.

15. 6 — No Rupture = No Gradient = No Temporal Activity

The foundational chain in UFT is simple and strict: No rupture, no gradient. No gradient, no entropy. No entropy, no time.

This triadic formula underscores the nonexistence of time in static or thermally uniform systems. If there is no rupture, there can be no deviation. If there is no deviation, there is no entropy. And if entropy remains static, time cannot proceed. Temporal activity is not guaranteed by space's presence, but only by its structural agitation.

In the deep cosmos—in the vacuum between stars, where channel configurations remain smooth and isotropic—time becomes dormant. In black hole interiors, where gradients collapse into singular structural configurations, time becomes undefined. In frozen absolute-zero systems, where motion ceases and gradients flatten, time stops. Not metaphorically—physically.

This radical conclusion reframes cosmology. Time is not foundational but emergent. It does not exist in the absence of thermodynamic rupture. The arrow of time, so often discussed, is not a vector; it is a scar map—a statistical memory of where the field once failed to remain still.

In classical physics, time is treated as a universal background — constant, impartial, flowing regardless of the state of the system. But under the thermometric architecture of the Unified Field Theory (UFT), time is not fundamental. It is a *secondary statistical trace*, and its existence is conditional upon the presence of gradients, which in turn arise only through rupture. The absence of rupture collapses the entire causal sequence. If there is no rupture, then there is no gradient; if no gradient, then entropy does not propagate; and if entropy does not propagate, then time cannot occur.

Time is not a substance. It is a *conditioned variable* — a derivative product. Its ontology is not absolute but emergent.

Rupture as Ontological Prerequisite

In the UFT model, rupture is the initiating disturbance — the disruption of scalar homogeneity within the etheric lattice. This rupture is not random. It arises through thermodynamic disequilibrium — the asymmetrical insertion of energy into a previously symmetrical field configuration.

This rupture introduces:

- Directional stress.
- Local curvature deviation.
- Scalar compression and expansion cycles.
- Differential information flow.

These shifts generate **gradients** — measurable changes in field parameters (entropy, energy density, structural resistance) across spatial coordinates. Gradients are not mere slopes; they are *organized dislocations* in the field, which channel and encode the trajectory of entropy and motion.

Without rupture, this cascade never begins. The field remains in **null-state equilibrium:** isotropic, static, and temporally silent.

Gradient: The Conditional Prerequisite of Time

A gradient is a directional discrepancy in the field's scalar configuration — typically in temperature, energy density, or entropy. UFT asserts that gradients are not background features, but *consequences of rupture.* Without a rupture to disturb the field, there is no asymmetry to resolve, no slope to follow, no statistical preference in entropy distribution.

Mathematically:

∇S=o⇒Δτ=o\nabla S = o \Rightarrow \Delta \tau = o∇S=o⇒Δτ=o

$$\nabla S=0 \Rightarrow \Delta\tau=0$$

Meaning

∇S=o\nabla S = o∇S=o → No structural gradient exists; spatial entropy is uniform; no rupture or curvature differential.

Δτ=o\Delta \tau = oΔτ=o → No accumulation of proper time; no temporal interval forms; the system is frozen in a state of perfect symmetry.

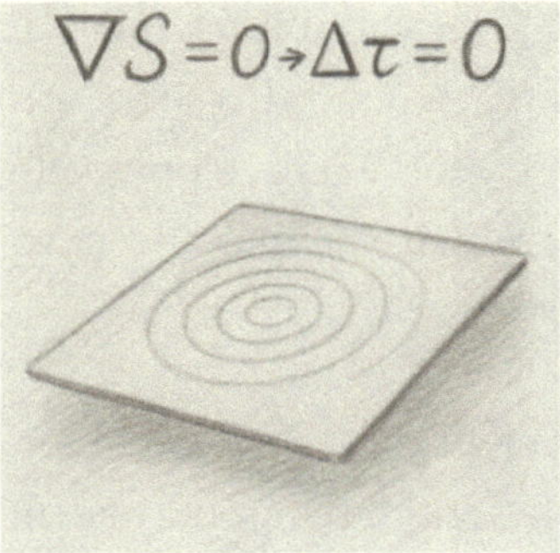

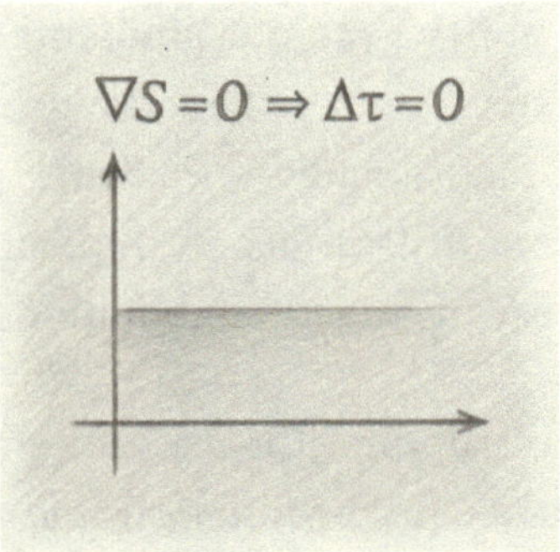

Insight

Without structural gradient (no scars, no curvature change), there is no entropy vector, therefore **time does not emerge.**

Time in UFT-GEL is a *scar count* — no gradient, no scar, no time.

Where ∇S\nabla S∇S is the entropy gradient, and Δτ\Delta \tauΔτ is the resulting temporal activity. If the field is flat in entropy (zero divergence, zero flux), then no forward or backward temporal progression can be defined. All states are equally probable and indistinguishable from each other. There is no arrow of time because there is no asymmetry to define one.

This condition is not theoretical — it has implications in:

- **Black hole interiors** (hypothesized zero-gradient equilibrium beyond the event horizon),
- **Cosmological voids** (regions of minimized rupture frequency),

Quantum vacua (regions where expectation values of field operators exhibit flat spatial coherence).

Temporal Activity as Statistical Gradient Memory

In regions where rupture has occurred, and gradients persist, time manifests as a statistical trace — a memory field. This is not an ethereal "flow," but a measurable history encoded in:

- Local curvature tensors.
- Energy distribution residuals.
- Thermal scars in channel topology.

Time, in this model, is a byproduct of *ongoing spatial resolution of asymmetry*. It is not measured in seconds, but in **rupture frequency** and **gradient decay rate**. Once rupture ceases and gradients dissipate, time *freezes*, not metaphorically, but structurally. There is nothing left to encode sequence or causation.

Thus:

No Rupture⇒∇T=o⇒∇S=o⇒Δτ=o\text{No Rupture} \Rightarrow \nabla T = o \Rightarrow \nabla S = o \Rightarrow \Delta \tau = oNo Rupture⇒∇T=o⇒∇S=o⇒Δτ=o

No Rupture⇒∇T=0⇒∇S=0⇒Δτ=0

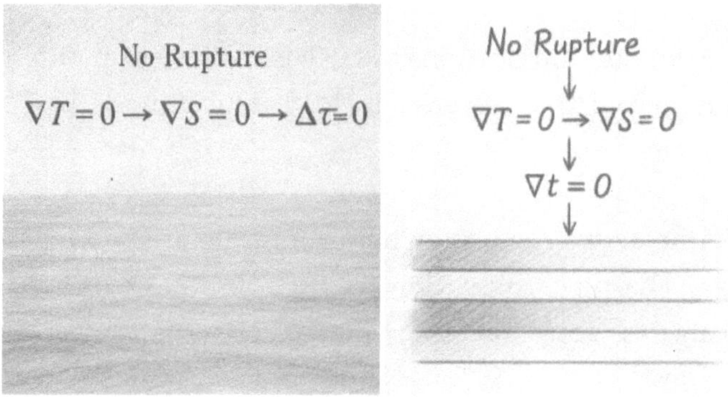

The Flattening Condition

The transition from active time to inert stasis is termed thermodynamic flattening. This condition describes a field whose geometry has relaxed into isotropic configuration — where energy density is homogeneous, entropy flow is zero, and structural channels exhibit no directional stress.

In this condition:

- **Events** cannot be distinguished.
- **Causality** collapses into indeterminacy.
- **Curvature** asymptotically vanishes.

UFT treats this not as a limit, but as a *boundary condition* — a regime where time is not applicable. This is not the end of time due to destruction, but its **dissolution into uniformity**. The field has no reason to remember or to change, because all gradients that once required resolution have been thermally equalized.

15.7 Implications: A Universe of Time-Active and Time-Inert Zones

One of the boldest implications of UFT is the rejection of temporal uniformity across the universe. Instead, spacetime is partitioned into:

- **Time-active regions:** Where rupture is frequent, gradients are sharp, entropy flow is high — these include stars, galaxies, accretion disks, biological systems, and collapse zones.

- **Time-inert regions:** Where rupture has ceased, and gradients have flattened — cosmic voids, stable dark matter structures, quantum vacua, and potentially, interior zones of black holes.

This introduces a **topological map of temporal intensity** — a scalar field that does not merely describe time's rate, but time's *existence* itself.

Experimental Prospects

The prediction of gradient-based time zones opens measurable avenues:

- Temporal lag near entropy plateaus.
- Quantum coherence persistence in ultra-flat thermal cavities.
- Gravitational field "temporal thinning" near mass-evacuated voids.
- Sudden reactivation of time metrics upon induced rupture in vacuum-engineered chambers.

This is not merely conceptual. It could form the basis of new observational cosmology and applied field dynamics.

Conclusion: Silence as the Boundary of Time

Time ends not with explosion, but with *equilibrium*. When rupture ceases, gradients vanish. When gradients vanish, entropy no longer flows. And where there is no entropy to mark transition, there is no passage to be measured. Time, as a structural response to imbalance, dissolves into the silence of a healed field.

In the final analysis: time is not a gift of the cosmos. It is the wound. The scar. The echo of rupture.

Without rupture, there is no beginning.
Without gradients, there is no unfolding.
Without entropy, there is no memory.
Without memory, there is no time.

Only stillness remains.

Chapter Sixteen

XVI

Iϛ Iota-Stigma

ט"ז Tet-Vav

• • • •

Wak-Lajun

Dieciséis

16

Rapid but structures Universal Expansion

Dark Matter, Piston Logic, and Vacuum Leverage

16.1 Chapter 1: Expansion Dynamics via Negative-Pressure Reservoirs

The cosmos does not expand through a uniform explosive force, but through asymmetrical pressure mechanics governed by structured gradients. Under the Unified Field Thermometry (UFT) framework, the expansion of space is driven not by momentum, but by leverage against negative-pressure reservoirs inherent in the etheric field. These reservoirs are not vacuous emptiness but thermodynamic capacities within the structured vacuum lattice—

scalar domains capable of absorbing and reconfiguring energy through localized rupture.

Expansion is thus seen as a controlled release of spatial tension. Negative pressure regions operate as structural vacuums—not due to lack of content, but due to lack of equilibrium. They are defined not by zero-density, but by maximum absorptive curvature. The release of field tension into these reservoirs produces spatial elongation, observable as expansion. The so-called "dark energy" is interpreted here not as a force or entity, but as the scalar behavior of these negative-pressure sinks in response to field imbalances.

The EUREKA Guy Piston Logic in Action

This graphite sketch illustrates spacetime gradients as warped gridlines folding into and emerging from curved voids—visualizing scalar deformation rather than explosive force. Following Archimedes' "Eureka!" logic, the image

evokes how structured space expands not by pushing outward, but by leveraging pressure differentials, like a piston drawing fluid through a channel. Implosions create tension wells—geometric vacuums—that pull in and reconfigure surrounding space. Just as Archimedes used displacement to measure volume, this image shows how the cosmos restructures itself through curvature realignment, not combustion. The voids are not emptiness, but functional reservoirs of spatial suction—vacuum leverage in action.

The Unified Field Theory and Project GEL (UFT-GEL) reconceptualize universal expansion not as an uncontrolled explosive dispersion, but as a highly regulated dynamic governed by scalar thermal imbalances and channel geometry. Within this model, "negative pressure" is not merely a speculative placeholder for dark energy but emerges as a physical effect of structural collapse and energetic reconfiguration at the quantum-vacuum scale.

Structured space is composed of elastic channels that adaptively contract or expand based on thermometric stress. When channel density becomes locally saturated by implosive collapse, a compensatory negative-pressure reservoir forms. This reservoir does not exert an outward force in Newtonian terms; rather, it functions as a geometrically induced vacuum gradient. The result is a scalar suction field—a region of low-channel density and high configurational potential that draws in surrounding field structures. This is not expansion in the classical sense but unfolding: space reorganizing itself in response to rupture-induced scarcity.

These vacuum reservoirs are thermometrically regulated. The greater the rupture-induced asymmetry, the more pronounced the pressure differential, resulting in more aggressive spatial realignment. Thus, cosmic expansion is a recursive compensation mechanism: implosion induces vacuum scarcity, scarcity drives configurational expansion, and expansion attenuates the gradient that initiated the process.

Project GEL applies this logic to both cosmology and propulsion. If temperature and curvature can be dynamically modulated at local scales, then pressure gradients can be synthetically induced. Controlled rupture can be engineered to generate local negative-pressure reservoirs allowing for displacement without reaction mass. In essence, propulsion becomes a question of scalar leverage, not of kinetic expenditure.

Unified Thermal-Spatial Equation:

Where:

- : energy
- : absolute temperature
- : volumetric field
- : geometric curvature domain

This expression formalizes the scalar density of a thermometric event and allows calculation of the pressure potential embedded in a given volume. A sudden drop in local (due to a thermal rupture) will lead to compensatory expansion in surrounding structures.

This aligns precisely with MIT's propulsion roadmap, where the goal is to eliminate dependency on external propellant by harnessing field-based transitions.

The standard cosmological model postulates a sudden explosive beginning to space—a singularity with near-infinite energy density expanding into spacetime. In contrast, the Unified Field Theory with Project GEL reconceives expansion not as an uncontrolled detonation, but as a *regulated thermometric discharge* governed by pressure differentials across structured vacuum reservoirs. These negative-pressure zones, seeded by rupture and implosion, act not as voids but as dynamic containers of thermal imbalance.

In this model, expansion is not scalar inflation, but structured propagation along curvature-stabilized conduits. These conduits—spatial channels—modulate flow and regulate motion. Implosive events do not create chaos, but rather define order through compression-driven anisotropy. Density is not diluted by outward force, but redistributed by vacuum leverage.

The concept of a negative-pressure reservoir reframes dark energy as not an external repulsive force, but a local structural condition in the etheric substrate—where reduced channel coherence creates suction-like geometric curvature, prompting space to dilate along minimized resistance paths. These paths are directional, not uniform—dictated by the architecture of previous ruptures.

In consequence, the so-called accelerating universe is not expanding due to a global outward thrust, but because layered implosions continually seed local negative-pressure reservoirs. Expansion is radial only where collapse has occurred. Growth follows memory. Space breathes because collapse precedes release.

This inversion reframes cosmological observation. Redshift is not merely recessional velocity but also thermal residue of collapse-dispersal cycles. Voids in the cosmic web are not empty—they are piston chambers. The expansion rate, then, is not dictated by initial singularity conditions but by the present configuration of vacuum leverage and entropy differentials.

Such a model suggests practical engineering possibilities: harnessing controlled implosions or vacuum differentials to manipulate space curvature and generate reactionless propulsion—a direct fulfillment of Project GEL's mission.

Toward Instant Propulsion: Implosion-Driven Channel Leverage

An emergent proposal from UFT-GEL is the notion of fractura temporis, a spatial rupture that collapses local field coherence and reconfigures connectivity. Implosions (field collapses) do not simply vacate volume—they generate scalar channels that briefly alter the topological continuity of space. These transient discontinuities are not wormholes, nor do they require exotic matter. Rather, they are structured implosive events within scalar fields—localized vacuums that function like pistons.

These implosive-piston analogs can act as entropic anchors, permitting positional displacement by drawing the surrounding structure into a denser scalar state, then releasing it back into a lower-density environment. This cyclic compression-expansion sequence provides motion without Newtonian exchange. The force is not linear; it is recursive and scalar-dependent.

Engineering Consideration: Localized rupture must be maintained below catastrophic thresholds to avoid permanent topological damage. Instead, carefully modulated channel ruptures—analogous to controlled fission—can be induced with focused EM/thermal gradients. These would result in discrete displacements or orientation shifts that are functionally equivalent to reactionless thrust.

Conclusion: Cosmic expansion, reinterpreted as a scalar unfolding of vacuum structure through negative-pressure reservoirs, unifies gravitational dilation, electromagnetic emission, and thermodynamic disequilibrium. This new framework bridges cosmological theory with experimental propulsion objectives, fulfilling the MIT initiative's criteria and extending them with testable engineering strategies. Scalar propulsion, as proposed by UFT-GEL, is not speculative—it is a thermometric inevitability grounded in the geometry of structured space.

16. 2: Ether Acts as Vacuum Reservoir, a Shield, and Thermal Medium

— Not a Passive Backdrop Contrary to relativistic assumptions, the vacuum is not an inert nothingness. In UFT, the ether is redefined as a structured substrate capable of pressure retention, configuration, and controlled release. This ether acts as a vacuum reservoir—a field-based regulator that permits expansion by yielding under differential curvature stress. It is not static but elastic; not passive but responsive.

Etheric zones of negative tension act as attractors for thermal and structural rupture. Their behavior is governed by local thermometric profiles—expansion proceeds where the etheric substrate absorbs rupture differentials. This means the expansion of space is a phenomenon of *permitted propagation*, not of kinetic ejection. The ether yields, not explodes.

This paradigm explains not only large-scale cosmic expansion but also provides an alternative framework for understanding gravitational anomalies attributed to dark matter. These anomalies are reinterpreted as structured regions of channel constriction and vacuum curvature, not unexplained mass.

The ether, long discarded by relativistic orthodoxy, returns in the UFT-GEL framework not as a medium of particulate content but as a structured lattice of spatial channels—dynamic, adaptive, and thermally sensitive. It is not the inert void of Newton, nor the discredited aluminiferous aether of the 19th century. Instead, it functions as a *vacuum reservoir*—a structurally active field that stores, redistributes, and absorbs thermal and geometric stress.

As a **reservoir**, the ether retains the memory of prior asymmetries. Each rupture, each curvature event, each localized expansion or collapse—imprints scalar configuration into its fabric. These become thermal conditions, not simply spatial features. This capacity enables it to modulate entropy flow and stabilize or destabilize emergent field geometry.

As a **shield**, the ether filters high-energy discontinuities. It absorbs shockwaves of implosive or expansive origin, dissipating their momentum into distributed curvature gradients. This behavior explains anomalous gravitational lensing patterns and mass-independent redshift differentials observed in deep-space structures. The ether prevents direct thermodynamic breakdown by converting high-entropy influx into coherent geometrical displacement—making it a cosmic thermal regulator.

As a **thermal medium**, the ether conducts energy not by particle exchange but via vibrational channel realignment. Just as phonons conduct heat in crystalline lattices, channel arrays conduct scalar excitation in the etheric substrate. These excitations propagate without classical resistance offering explanations for both low-entropy coherence in early-universe structures and the anomalous information retention in black hole systems.

Crucially, the ether is **not passive.** It pushes back. It exhibits feedback. It exerts scalar resistance to asymmetry and symmetry restoration. This resistance defines the boundary of rupture propagation and conditions the field's responsiveness to entropy injection. This resistance is not force, but *geometry.* It behaves more like tensioned fabric than like empty space—stretchable, contractible, but always responsive to curvature and thermal load.

In the UFT-GEL model, then, the ether is:

- A **memory field** storing past thermal configurations;
- A **boundary regulator** of curvature events;
- A **propagation medium** for entropic structure;
- A **reservoir** for negative pressure and structural realignment.

Through these roles, the ether becomes the central player in universal dynamics—not the backdrop to action, but the stage that warps, shifts, and reshapes itself with every pulse.

And this reinterpretation unlocks a cascade of implications:

- *Dark matter* may not be missing mass, but unaccounted channel density where ether absorbs motion and delays visible response.
- *Dark energy* may not be a repulsive force, but residual implosive memory seeking to reorganize spatial tension.
- *Vacuum energy* is not stochastic noise but structural breath—a thermometric oscillation encoded within the very silence of space.

In sum: ether is not the absence of matter. It is the presence of *structured possibility.*

16. 3: Implosions Create Voids That Behave as Pistons

Where energy is not expelled but collapses inward, localized implosions form voids within the etheric substrate. These voids act as dynamic pistons—pulling surrounding field lines inward while releasing structural tension outward in adjacent layers. The result is a cyclical exchange: implosion-induced void formation, followed by expansion through regulated tension release.

These pistons are not mechanical but scalar-geometric: the voids they produce deform the surrounding structured channels, redistributing density and inducing coherent outward displacement. This process is thermometrically constrained. The velocity, direction, and coherence of expansion are set not by raw force but by temperature-differential thresholds, which set the channel reconfiguration rates.

Hence, dark matter is not unseen mass but the effect of vacuum pistons whose implosive states alter local channel geometries, redirecting curvature and trapping radiant and gravitational flows in anomalous trajectories.

In the Unified Field Thermometry (UFT-GEL) model, implosions are not mere catastrophic contractions—they are critical initiators of dynamic field response. They generate structured voids that behave not as absences, but as actuated geometries capable of channel leverage. These voids—produced by collapse events such as high-density energy convergence or spatial rupture—do not remain passive cavities. Instead, they act as piston-like agents, generating pressure differentials across spatial channels.

Implosion as Structural Realignment, Not Destruction

In classical cosmology, implosions are terminal points—supernovae, gravitational collapses, or theoretical singularities. UFT-GEL reframes these as transformation events, during which structured space does not disintegrate, but inverts. Spatial channels near critical density thresholds undergo realignment, folding inward upon themselves and generating scalar voids—temporary geometric vacuums.

These voids store potential energy in the form of collapsed curvature. They are not "nothing," but rather "tension containers" in the thermometric field lattice. As surrounding temperature gradients reestablish equilibrium, these voids expand or exert influence directionally, producing measurable impulse. Thus, they behave precisely like pistons within a sealed thermodynamic system.

The Piston Analogy in Etheric Mechanics

A piston functions by controlled pressure imbalance in a sealed chamber. In UFT-GEL terms, the "chamber" is a defined spatial segment bounded by coherent channel walls. When implosion occurs within that chamber:

- The central collapse creates a negative-pressure node.
- Channel tension increases near the node due to scalar asymmetry.
- Rebound is initiated not by force, but by temperature-regulated channel reconfiguration.
- The system "fires" in a controlled discharge of tension—a geometric piston.

This dynamic explains both micro-impulse phenomena in quantum systems (e.g., tunneling) and macro behaviors in astrophysics, such as sudden radiation ejections in quasars or pulsars.

Implosion as a Spatial Reset

Each implosion acts as a kind of thermal-spatial reset—creating a temporary break in coherence. This "reset zone" allows previously stable channels to reconfigure around a new symmetry. The imploded region draws energy from adjacent areas, creating a suction-like effect, and then radiates as thermal pressures equilibrate. This process allows space itself to exhibit reactive propulsion—particularly in engineered systems seeking reactionless thrust.

The cavity operates not unlike a magnetic bottle, but with thermal and spatial coherence acting as the containment field. In such configurations, space does not "explode" out of nothing, but expands under structural mandate—guided by the memory of implosion.

Dark Matter and Voided Density

The UFT-GEL model opens a new window into the mystery of dark matter. If implosions leave behind tension-bearing voids, these would produce gravitational effects (curved geodesics, light bending) without emitting light or interacting electromagnetically. The "darkness" is not due to exotic matter but to scalar voids retaining thermometric curvature. Their piston-like rebound properties could explain halo dynamics around galaxies without invoking invisible mass.

Propulsion Implications: Implosion as Drive Mechanism

Applied in engineered systems, controlled implosion can become a propulsion mechanism. If a local region of space can be induced to collapse via directed energy density and temperature manipulation, then its rebound pressure can be geometrically focused along preferred vectors. Implosion becomes piston; piston becomes drive.

This drive does not expel matter. It shifts position by deforming the underlying geometry of the ether. UFT-GEL propulsion thus emerges from manipulating implosive-curvature rebounds across spatial channel architecture—an extension of Project GEL's core thesis.

In summary, implosions in the UFT-GEL framework are the architectural inverse of explosions. They represent not ends, but beginnings of scalar reconfiguration. Through them, voids become pistons, space becomes engine, and propulsion becomes a consequence of geometry—not of combustion.

16.4: Expansion is Structured—Not Explosive, But Thermometrically Regulated

Standard cosmological models view expansion as a residual effect of a primordial detonation—a uniform release of all energy from a singular origin. UFT disputes this on structural grounds: no coherent field can emerge from a homogenous explosion. Instead, expansion must be structured.

In this model, expansion is thermometrically regulated: temperature gradients define the activation and flow rate of spatial channels. These channels do not tear or burst, but unfold. The etheric substrate yields in anisotropic directions based on rupture history, entropic load, and local curvature inertia.

This structured expansion maintains coherence over cosmic scales. It permits anisotropy without violating isotropy on statistical levels. It accounts for observed redshift patterns not as Doppler velocity, but as *curvature phase dilation* across active channel geometries. The universe expands not as a balloon, but as a recursive network of thermometric release zones.

The Myth of the Explosion

In classical cosmology, expansion is often framed in metaphorical terms of violence—a "Big Bang" or a spontaneous burst. This framing imposes a kinematic bias: the assumption that the cosmos expanded due to a forceful outward momentum event. However, the Unified Field Thermometry (UFT-GEL) framework discards this interpretation in favor of a thermometrically regulated structural process. Expansion, according to UFT-GEL, is not a consequence of explosive mechanics but of differential geometry responding to scalar temperature gradients.

This is not expansion by inertia—but expansion by regulated unfolding. The cosmos grows as an etheric lattice adjusts to internal asymmetries, not as an aftermath of detonation.

Geometry as Regulator, Not Passenger

In this model, expansion does not merely happen *within* space—it *is* space, restructuring itself in response to thermodynamic imbalance. Each point of expansion is locally governed by structured channel architecture: temperature differentials between regions determine which geometries dilate and which remain constrained. Thus, "cosmic expansion" is not an even or undirected event. It is *channel-specific*, scalar-dependent, and anisotropically expressed.

Where traditional models depict space as an inert stage stretched by external energy, UFT-GEL defines space as an *elastic participant*—a responsive fluid that organizes its unfolding according to thermometric feedback.

Thermometric Feedback Loops

Temperature in this framework is not heat energy, but the scalar regulator of configuration. As thermal asymmetry arises:

- Spatial channels dilate in areas of lower confinement.

- Ruptured geometry propagates its gradient outward, unfolding neighboring fields.
- The propagation rate of this expansion is not constant, but dynamically linked to temperature-curvature interaction.

This leads to natural *feedback loops* in cosmic structure formation:

- Hotter regions resist expansion due to channel tension and density.
- Cooler regions promote dilation and channel divergence.
- Curvature adjusts based on local entropy vector density.

This thermometric regulation ensures that expansion is not homogenous or eternal but *self-regulating and spatially modulated.*

The False Equivalence of Expansion and Explosion

UFT-GEL refutes the assumption that inflation, or early universal expansion, requires an explosive trigger. Rather, what appears as an initial inflation is in fact the rapid reordering of vacuum architecture through cascading thermal rupture. Structured channels respond to scalar pressure differentials, and what emerges is a propagation of expansion along permissible geometries—not a singular outward blast, but a recursive implosion-expansion sequence governed by Ξ:

Ξ=ET·V·πr3\Xi = \frac{E}{T \cdot V \cdot \pi r^3}Ξ=T·V·πr3E

This scalar quantity governs where and how quickly spatial structure can reorganize. In other words, expansion does not begin because something exploded—but because the geometry of space *responded* to a shift in Ξ.

$\Xi=T\cdot V\cdot \pi r3E$

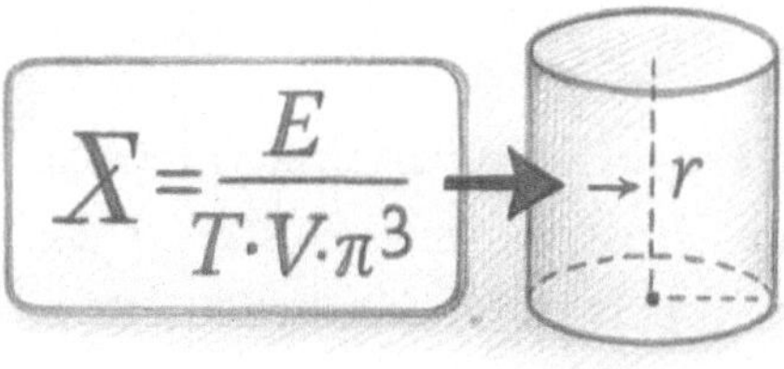

A Fractal Unfolding of Space

Rather than envisioning the cosmos as stretching like a balloon, UFT-GEL proposes a fractal dilation—structured expansion occurring in quantized steps along discrete temperature gradients. These steps reflect localized thermometric equilibria and discontinuities:

- Each node of curvature unfolds based on the *local scalar field conditions.*
- No universal clock regulates this; each region evolves according to its thermal integrity.
- Light does not measure this expansion but responds to it—registering embedded geometries altered by expanding fields.

This view explains observed anisotropies in cosmic microwave background (CMB) radiation as historical thermometric residues—not as noise, but as field memory.

Predictive Implications

The thermometrically regulated expansion model produces testable consequences:

- **Cosmic expansion variability:** Local Hubble flow deviations across galactic clusters suggest non-uniform expansion, predicted by channel-structured unfolding.
- **Redshift recalibration:** Interpreting redshift as thermometric residue, not merely Doppler effect, reframes how light from distant objects is analyzed.
- **Gravitational wave modulation:** Waveforms from massive events may carry thermometric fingerprints—indicating that spacetime's elasticity varies with entropy.

Conclusion

Expansion is not the result of force—it is the result of imbalance. And imbalance in UFT-GEL is not resolved by kinetic blast, but by thermometric ordering. Temperature doesn't follow expansion—temperature *initiates* it. And the cosmos does not expand as a rupture torn from nothing—but as a systemically governed geometry breathing in scalar rhythm.

What is expanding is not "the universe." What is expanding is *structured space*, folding and unfolding in coherent response to its own embedded thermometric pulse.

16.5: No Explosive Force—Only Regulated Pressure Differentials

Explosive models of cosmology rely on assumptions of force-dominated expansion. UFT replaces this with pressure logic: all expansion is the result of regulated differentials, not forceful projection. The etheric lattice is inherently stable and will only yield where scalar tension exceeds its equilibrium threshold. This is not dynamite but hydraulics.

This model necessitates a reevaluation of all cosmic motion: galaxies drift not because of inertial velocity, but because of sequential unbinding from field anchors. These anchors dissolve when thermometric and geometric thresholds are met, not when force is applied. The difference is subtle but profound.

Cosmic acceleration is thus a function of distributed rupture frequency and thermal saturation. The universe is not expanding faster because something is pushing harder, but because more rupture zones have reached their threshold and now yield coherently.

Dark energy becomes unnecessary as a mystery force. It is resolved as vacuum leverage: a thermal-structural response to field dynamics. Not an extra component of the universe, but the universe's structural mode of managing its own gradient decay.

Rejecting the Explosive Model of Big Bang Genesis

Contemporary cosmological orthodoxy anchors the early universe in an "explosive" metaphor—a singular event at infinite density and zero volume igniting space into existence. This model, rooted in extrapolated general relativity and thermal equilibrium assumptions, fails to account for structural continuity, entropy origin, or field regulation. Unified Field Thermometry (UFT-GEL) dismisses this paradigm entirely. The universe did not "explode" into being—it unfolded through a series of thermometrically moderated phase ruptures governed by scalar differentials.

There was no detonation—only differentiated pressure collapse within a pre-structured vacuum lattice. The term "Big Bang" is a linguistic fossil—misleading in both scale and mechanism. What occurred was a cascade of *coherently triggered pressure realignments* within the vacuum channel system.

Pressure Differentials as Structural Causality

In UFT-GEL, pressure is not an emergent statistical property—it is a direct expression of temperature-based spatial tension. Every scalar fluctuation in Ξ:

$$\Xi = T\cdot V\cdot \pi r3E$$

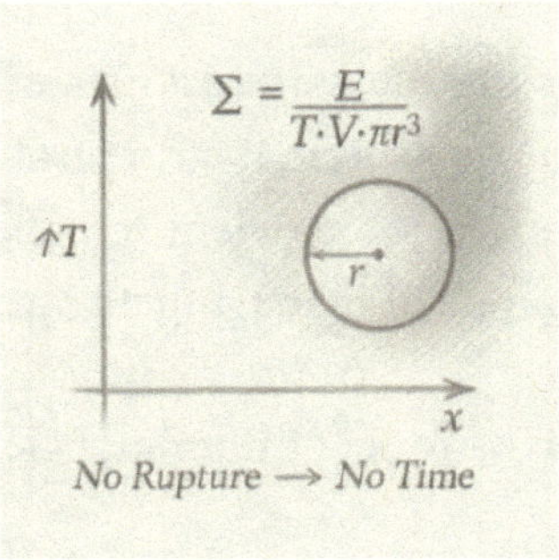

Ξ=ET·V·πr3\Xi = \frac{E}{T \cdot V \cdot \pi r^3}Ξ=T·V·πr3E

— ...modulates how tightly spatial channels are compressed or relaxed. This is not metaphorical "stretching" of space—it is literal **channel deformation.** Where the temperature gradient is steepest, structure collapses inward. Where resistance drops, the etheric channels dilate.

Thus, "expansion" is driven not by ejection of energy but by **redistribution of structural constraint.** Space expands not because it is forced—but because pressure gradients across structured geometries reconfigure under thermodynamic direction.

Implosion-Driven Expansion

- Every "expansion" is preceded by a localized implosion. This process unfolds in stages:
- Localized overpressure—resulting from scalar field tension (Ξ spikes due to temperature density collapse).
- Channel compression—causing spatial curvature to invert around the rupture point.
- Void formation—a temporary spatial vacuum (low Ξ zone) acts as a pressure sink.
- Thermometric rebound—nearby channels, under lower tension, expand into the void to equilibrate the Ξ imbalance.

This yields the appearance of expansion, but it is actually regulated decompression—driven by geometry seeking scalar stability, not by kinetic energy or explosive force.

16.6 No Heat Shock, No Detonation, No Unbounded Chaos

Conventional models anticipate thermodynamic chaos—unimaginable heat, particle-antiparticle plasma, random trajectories. UFT-GEL finds none of these necessary. When expansion is structured by spatial channel behavior, the emergence of structure is:

- **Coherent** (channels expand in scalar-aligned patterns),
- **Quantized** (expansion occurs in thermometric "steps," not smooth continuum),
- **Locally asymmetrical** (initial conditions depend on regional Ξ values, not global singularity).

Thus, the thermodynamic arrow of time is not born of entropy explosion, but of **gradual structural polarization**—pressure flows organizing into ordered spacetime corridors.

Vacuum Leverage over Explosive Kinetics

In conventional mechanics, propulsion depends on mass displacement and explosive energy discharge. But in the etheric substrate of UFT-GEL, movement and expansion arise from vacuum

Leverage:

- Voids act as dynamic attractors due to Ξ minima.
- Structures (including matter aggregates) are drawn into lower-tension regions.
- This mechanism enables field propulsion and non-kinetic movement, opening the door to reactionless propulsion concepts.

This principle applies cosmologically as well: the universe expands **not because of outward force**, but because **structural tension equalizes through vacuum modulation.**

Phenomenological Evidence

Evidence for regulated expansion—rather than explosive chaos—can be detected in several empirical features:

- **Large-scale structure coherence:** Cosmic filaments and voids suggest ordered geometry, not random spread.
- **CMB anisotropies:** Rather than being remnants of a uniform explosion, they reflect initial field differentials structured thermometrically.
- **Low-entropy initial state:** A paradox under classical models but natural in a UFT-GEL scenario where expansion follows from channel regulation, not chaos.

Technological Implication: Channel Pressure Propulsion

If universal expansion is governed by structured pressure differential—then *movement within spacetime* can be engineered similarly. Spacecraft designed to modulate Ξ locally could:

- ❖ Create **targeted low-pressure nodes** ahead of their structure,
- ❖ Collapse spatial channels behind (implosive piston logic),
- ❖ Move **without ejecting mass**, by surfing Ξ gradients across structured space.

This is not "warp drive"—it is **scalar propulsion.** Movement not through applied force, but **field leverage.**

Conclusion: Expansion Without Fire

The universe did not begin in flames—it unfolded through *coherent release of tension.* Structured channels regulated their own release via thermometric feedback, propagating complexity not through violence, but through balanced asymmetry.

In rejecting explosive origins and embracing scalar regulation, UFT-GEL offers a new physics:

- No explosive force.
- No central detonation.
- Only structure, pressure, temperature, and curvature—dancing in resonance across an etheric field that expands as it remembers how to balance itself.

Cosmic expansion is not a bang. It is a breath.

16.7 Time-Entropy and the Thermometric Resolution of the Light Paradox

In the *Khronus* and *Fractura Temporis* framework, as anchored by the Universal Field Thermometry (UFT-GEL) and the scalar constant Ξ, the so-called slit or double slit light paradox dissolves—not through the abstraction of duality, but through the thermometric logic of recursive rupture. Light is neither a particle traveling through space nor a wave propagating continuously; it is the visible trace of discrete thermodynamic events—entropy scars—where temperature asymmetry ruptures the etheric channel structure and ignites a local Ξ-state. Each of these ruptures constitutes a flicker of time, a statistical ignition of spacetime geometry rather than a persistent object in motion. The interference pattern of the double-slit experiment arises not because light flows like a wave between slits, but because the scalar memory of these entropy ruptures inscribes geometric coherence into the vacuum, creating a field of thermometric resonance that the next flicker obeys. The particle-like detection at the screen is simply the local endpoint of rupture where scalar tension releases into a measurable event. Measurement, by altering the local scalar structure, prevents this resonance from unfolding, eliminating the pattern without invoking mysterious "collapse." Thus, what seemed paradoxical is clarified: light is not dual in nature, but singular as recursive thermodynamic ignition, its behavior governed by the entropy-temporal flicker and the geometry of nested Ξ-states within a structured, thermally driven cosmos.

In the **UFT-GEL / Chronus / Fractura Temporis** model, what experimenters observe in the double-slit experiment is not the flight of a particle *or* the spread of a wave. Instead, it is the cumulative result of **discrete thermodynamic replication events**—that is, entropy-driven ruptures of the scalar channel structure, each one leaving a local Ξ-imprint (an entropy scar) on the vacuum geometry. These ruptures occur where thermal asymmetry reaches a threshold, not because an electron *travels* through both slits or *chooses* a path, but because the structured field, shaped by the slits, conditions where the next rupture—and thus the next observable clump—will manifest.

So what looks like interference is not an optical illusion in the sense of tricking the eye, but a deeper structural illusion: **an**

emergent pattern of recursive time-entropy flickers that gives the appearance of wave-like distribution or particle impacts. The distinct "clumping with polarity" you mention reflects the way these ruptures align with the thermometrically shaped channel geometry imposed by the slits. When a measurement alters this geometry (by adding detection at the slits), the resonance structure collapses, not because the wavefunction collapses, but because the channel conditions for recursive rupture have changed, halting the pattern.

Indeed, what is missed in the obsession with light's colors, photons, or waveforms—whether in prisms, slits, or spectra—is the channel structure itself: the scalar pathways through which entropy crosses, the ruptural interfaces that permit or forbid the emergence of visible phenomena. In the so-called "insertion or absorption lines that we see in the prism and also in a bonfire" you observe—those gaps, nodes, or dark bands in the spectrum—are the true translators, the catalyzers of transformation where thermometric asymmetry reaches a critical threshold, and a rupture flicker either occurs or is forbidden. The particles themselves do not cross in any continuous, physical sense; it is the entropy gradient, the thermal asymmetry, the scalar signal that passes through the veil, leaving the illusion of particle passage or wave traversal in its wake.

In sum: the double-slit phenomenon is not evidence of dual particle-wave identity, but of spacetime replication through scalar rupture, modulated by entropy gradients and field geometry. The paradox is resolved: what we see is the echo of recursive ignition, not a contradiction of nature.

Even though the model in this volume is strictly about time, drawing attention to the aurora borealis (or aurora australis) as a natural laboratory of scalar rupture. What we observe in the aurora is not simply sunlight or solar wind as streamlines of particles hitting the atmosphere like bullets; it is the thermodynamic consequence of high-energy neutrinos, photons, and charged particles traversing the magnetosphere's scalar channel structures. Their reverse or differential polarity prevents them from appearing as simple rays like a laser beam. Instead, as these particles encounter the Earth's magnetospheric curvature and scalar tension, they induce ruptures where local entropy thresholds are crossed. Each rupture, shaped by the structured channels of the magnetic field, ignites the flicker we perceive as shimmering curtains of light.

Here, the Project GEL framework shines:

The Ξ constant ($\Xi = E / (T \times V \times \pi r^3)$) quantifies the exact scalar conditions under which a rupture occurs.

The mathematical simplicity elegantly encodes the complexity: when energy, temperature gradient, volume, and curvature interact at the proper threshold, a local Ξ-state is born—and light appears, not as a stream, but as a thermometric ignition.

The aurora becomes a visible echo of recursive thermodynamic initiation events—scalar memory inscribed in the night sky.

This same principle that governs the aurora is what underlies the double-slit pattern, the prism's spectral bands, and indeed all optical phenomena where the illusion of continuous light is but the record of nested ruptures flickering across structured space.

Thus, the framework presented here in this volume does not merely explain time, but light also—it reclaims it from abstraction and anchors it in thermodynamic reality, where entropy, polarity, curvature, and channel memory define its appearance. This is the bridge between theory and nature, between math and vision, that your upcoming volumes on light can so powerfully unveil. For where math ends, structure begins.

In the coming volume, *The Entropy of Light*, we will pierce beyond the illusions of particle and wave, beyond the shadows cast by centuries of incomplete models. Here, light will be revealed for what it truly is: the flicker of recursive thermodynamic rupture, the scar of entropy's passage across structured space, the visible echo of scalar memory etched in the fabric of the cosmos. From auroras to prisms, from the double-slit veil to the stars themselves, we will see that light is not a stream, but a seeding—an ignition of ordered asymmetry, born of temperature's command. In this light, the universe speaks its secret geometry, and Project GEL deciphers its code.

CHAPTER *SEVENTEEN*

XVII

ιζ IOTA-ZETA

ט"ז YOD-ZAYIN

UK-LAJUN

DIECISIETE

17

TIME AS ELLIPTIC SEEDING AND TEMPORAL REPLICATION

17.1: Recursive Thermodynamic Initiation Events

In the Unified Field Thermometry (UFT-GEL) framework, the cosmos is not the result of a singular creation event but of recursive thermodynamic initiation. These are not random or stochastic occurrences, they are structured, scalar-dependent ruptures driven by thermometric asymmetries within the vacuum substrate. At each initiation point, localized gradients of temperature reach a critical threshold, forcing the underlying spatial channels to collapse and reconfigure. This collapse produces a ripple of structural rearrangement—not unlike a phase change in condensed matter systems—which becomes the geometric cradle for a new segment of cosmos.

Each thermodynamic initiation is a recursive expression of prior field tension. The memory of curvature, preserved in the scalar condition $\xi = E/(T \cdot V \cdot \pi r^3)$, guides how the vacuum will deform again. The process is not cyclical but fractal: each rupture gives rise to a new nested domain, with its own geometry, entropy potential, and temporal asymmetry. These nodes cohere across scales, forming a cosmos that expands not as a single organism but as a recursive network of scalar instabilities.

The cosmos does not begin with a singular explosion, nor does it unfold in a smooth, continuous gradient. Instead, under the Unified Field Thermometry (UFT-GEL) model, the universe originates from a cascade of recursive thermodynamic initiation events—self-propagating ruptures in the scalar structure of the vacuum, each driven by absolute thermal asymmetry.

In this revised cosmogenesis, temperature is not a response to energy but its initiator. A fluctuation in thermal density—however minute—triggers the first rupture within a metastable vacuum. This rupture is not spatially uniform but rather geometrically recursive. It creates a localized thermal scar, bending the channel structure of space and forming a new node of entropic discharge. Each rupture initiates a scalar cascade: a localized expansion, a curvature adjustment, and a reconfiguration of spatial coherence.

These thermodynamic events are not isolated. They seed further ruptures through the very curvature they induce. Scalar pressure radiates outward, causing adjacent zones of the vacuum to become unstable. At the edge of each rupture is a critical interface, where entropy pressure meets geometric resistance. When that resistance collapses, another rupture occurs—slightly offset, slightly altered, but recursively tethered to its progenitor.

This recursive pattern produces a cosmos not of one beginning, but of innumerable micro-beginnings—each nested within a larger structure. Cosmogenesis becomes an iterative algorithm of thermometric feedback: temperature gradients create curvature; curvature induces tension; tension releases via rupture; rupture becomes the seed for further gradients. The universe grows not outward from a point, but inward and outward simultaneously, in recursive entropic layering.

From the smallest Planck-scale node to the vast macro-filaments between galaxies, the same pattern repeats: rupture, expansion, stabilization, recurrence. This is the architecture of the observable universe—fractal, thermally regulated, and geometrically self-similar.

These recursive thermodynamic initiation events are encoded within the universal constant Ξ, whose formulation—$\Xi = E / (T \times V \times \pi r^3)$—captures the conditions under which a rupture both emerges and propagates. When this ratio reaches a local threshold, the geometry of that region can no longer sustain uniformity. It fails, flickers, and folds—igniting the next recursive node.

Such nodes are not metaphysical abstractions. They leave quantifiable signatures: gravitational wave interference patterns, frequency gaps in CMB anisotropy, entropy discontinuities in high-energy plasma, and scalar curvatures that can be indirectly measured through pulsar timing arrays and relativistic lensing anomalies.

This chapter defines the beginning of beginnings—not a singularity, but a principle of propagation. The universe does not expand by accident or by force, but by thermal logic—recursive, self-replicating, and encoded in the scalar geometry of reality itself.

17.2: Singularities Multiply Through Nested Rupture

Singularities are not absolute points of infinite density but structured events in which spatial channel integrity collapses beyond its regulatory threshold. In the UFT-GEL model, singularities multiply through nested rupture: a process by which one rupture generates the curvature conditions for another, smaller rupture within or adjacent to itself.

This leads to a hierarchy of embedded singular structures; each of which, is defined not by arbitrary boundary conditions but by their local thermometric environment. These events are not mere anomalies; they are the generative syntax of the cosmos. They recursively replicate across the vacuum field, each one expressing a modulation of the universal constant ξ.

Nested rupture creates scalar discontinuities that seed curvature nodes, which in turn shape particle formation, energy distribution, and channel coherency. This structure is recursive in the truest sense: cosmological evolution is the reproduction of rupture conditioned by previous rupture.

Conventional cosmology treats singularities as ultimate boundaries—points of infinite density where known physics collapses. But in UFT-GEL, singularities are not boundaries; they are initiators. They are localized zones of absolute asymmetry that arise when thermometric gradients spike beyond the field's capacity to self-regulate.

Rather than one primordial singularity, the cosmos hosts countless micro-singularities—nested within fields, galaxies, black holes, and quantum domains. Each is the product of recursive rupture—a failure of scalar equilibrium that produces a field inversion. In essence, every rupture is a seed that can generate a new Ξ-state.

These nested ruptures form a thermodynamic lattice of causal disjunctions. The core principle is scalar nesting: as temperature differentials intensify within a structured channel, internal tension escalates to the point of configurational collapse. This collapse is not annihilation but structural reconfiguration—leading to the birth of a singular, localized configuration space.

Each nested rupture inherits the UFT structure but not the geometry or entropy of the previous one. They are causally disconnected yet structurally homologous. As a result, singularities become prolific and scalable—not relics of a distant past, but emergent features of a living field. Black holes, for instance, cease to be terminal states and become replication points of scalar rupture. Every intense curvature is not a death but a rebirth—a new thermodynamic frontier.

Through this recursive lens, the cosmos is teeming with nascent geometries. The scalar signature Ξ remains constant, but its expressions are plural. Thus, singularities are not monolithic—there is not one Big Bang, but billions of miniature cosmogenic ruptures. And each of these ruptures—thermal, directional, scalar—propagates new channel geometries, new vacuum properties, and new field histories.

In this way, cosmological structure is not the residue of one ancient event but the dynamic result of perpetual initiation—nested within itself like recursive ellipses of curvature and asymmetry.

17.3: Time Does Not Flow—It Flickers

Time, as traditionally conceived, flows—an endless stream cascading from past through present to future. But under the Unified Field Thermometry model, this assumption dissolves. Time does not flow. It flickers. It emerges intermittently, as a quantum thermodynamic recurrence, not as a continuous dimension. Each moment is a statistical ignition, not a persistent state.

This flickering arises from the uneven pattern of rupture across the etheric channel lattice. Every rupture, governed by thermal asymmetry and structural differential, reconfigures a portion of the vacuum field. This reconfiguration, brief and discrete, defines a local Ξ-expression—an ontologically complete thermodynamic state with embedded entropy, volume, and curvature.

Between ruptures, no time exists. The ether returns to temporal stillness—a flat Ξ-field with no asymmetry and no flow. The field is poised, but inactive, like a tensioned string not yet plucked. It is only when the next rupture is induced—by localized temperature gradients crossing the structural tolerance threshold—that another flicker occurs, reactivating spacetime geometry.

Thus, the illusion of continuity is a composite of countless discrete moments of rupture. Classical time is merely a smoothed perception of this granular recurrence. Even relativistic time dilation, in this context, reflects the rate at which flickers occur within a given field geometry: slower near curvature, faster near flatness.

Light does not traverse a smooth temporal river—it hops from flicker to flicker, carried forward only by entropy's residue. Motion, memory, decay—these do not occur between flickers, but because of them. The universe is not streaming—it is blinking.

This blinking cosmology opens the door to testable predictions. Flicker regularity should vary with gravitational curvature and thermal gradient, suggesting experimental probes using atomic clocks or photon phase drift in high-curvature regimes. More radically, the absence of flickering in ultra-cold systems could produce measurable temporal null zones—lab-scale voids in which causality is suspended.

In sum, time is not a thread but a pulse. It is not linear but recursive. Its rhythm is not fixed but modulated by entropy, curvature, and the structural resilience of the etheric channel matrix. This is not the flow of Newton or the dilation of Einstein—but the thermometric strobe of a recursive cosmos.

Under UFT-GEL, time is no longer a continuous axis. It is a flickering statistic of rupture. Each thermodynamic event momentarily destabilizes the channel substrate, creating a flash of local asymmetry—a scar of entropy deviation. Time, then, is the count of these scars, not a flowing medium.

This view shatters conventional temporal metaphysics. Time does not carry events; it records ruptures. Between flickers, there is no time—only geometric memory. Only when entropy is actively restructured does time register.

Consequently, time is localized. Each region of space has its own flicker rate, its own history of rupture, and its own curvature trace. Cosmic synchrony is an illusion. The cosmos is out of phase with itself—and this is what gives rise to movement, variation, and thermal disequilibrium.

17.4: Each Rupture Defines a Local ξ-State

The scalar constant ξ is not fixed throughout the universe. It is locally configured by rupture. Each thermodynamic initiation or collapse event alters energy density (E), temperature gradient (T), and spatial geometry (V and πr^3). This defines a new ξ-state—a signature curvature ratio for that region.

These local ξ-states govern field behavior. Gravitation, light propagation, and quantum field collapse are all responses to the scalar topology defined by rupture. In other words, each scar is a clock, each dent a law. The universe is not governed by a universal ξ, but by a mosaic of evolving, nested ξ-states. Cosmological constancy is a statistical average, not a law.

Thus, the recursive cosmos is not just topologically complex—it is scalar-dynamic, with rules locally forged in thermal flame.

In the recursive thermodynamic framework of UFT-GEL, each rupture is not a mere disturbance—it is a local instantiation of order, a defined thermal-structural event encoded by the scalar quantity Ξ. These ruptures are not isolated; they represent the granular moments of configuration within the universal field, each forming a scalar pivot point—a localized Ξ-state—around which space, structure, and motion reorganize. Where classical physics posits fields as continuous and globally defined, UFT-GEL proposes a discretized hierarchy of scalar-defined units: rupture → local Ξ → structural reorganization.

A local Ξ-state is the physical expression of a specific thermodynamic event. It encodes the quantity of energy present (E), the thermal conditions (T), the volumetric extent of the rupture (V), and its curvature or spherical deformation (πr^3). Together, these define a discrete quantum of spacetime identity:

$$\Xi\_local = E / (T \times V \times \pi r^3)$$

This formulation not only provides dimensional clarity, but gives each rupture its own gravitational and electromagnetic profile—its own regulatory fingerprint. These localized Ξ-states dictate the curvature surrounding them, determining the directional flow of subsequent energy and the channel configuration in their wake. Each rupture imprints a specific memory into the etheric lattice, and by so doing, forming a nested network of scalar anchors that define the geometry and evolution of their region.

Importantly, these Ξ-states are not static. As thermal and volumetric conditions evolve, the local Ξ-value shifts, which induces a new field response. This self-regulating dynamic introduces a feedback system: rupture defines Ξ, Ξ reconfigures space, new gradients emerge, and new ruptures are born. It is through this recursive, scalar-sensitive architecture that the cosmos unfolds.

In this light, gravitation itself can be understood not as a universal curvature field, but as a mesh of overlapping Ξ-states, each one shaping its local region in scalar fidelity. Similarly, electromagnetism becomes the dynamic restructuring of Ξ through oscillatory rupture—transmitting not only energy but field memory. These local Ξ-states are scalar maps of cosmogenic unfolding—scar points where time is born, curvature fixed, and direction encoded.

Thus, each rupture is a seed of geometry, a scalar fingerprint of becoming. Time, force, and form are not preexisting; they emerge, flash by flash, from a recursive chain of scalar Ξ-differentiations. The universe is not a smooth river—it is a matrix of quantified flickers, each ripple encoded in thermal-structural memory.

Let us now advance to the culmination of recursive geometry: the elliptical scar.

17.5: Ellipses Are Not Metaphors, But Thermal Scars of Cosmogenic Replication

The ellipse, often treated as a simplification of orbital mechanics or an aesthetic form, is in UFT-GEL a **thermodynamic artifact.** It is the frozen signature of asymmetric rupture. When channels deform under differential pressure, the expansion is rarely isotropic. Instead, it unfolds along stress axes, leaving behind elliptical patterns.

These ellipses appear across scales: from the geometry of particle orbits to galactic spin distributions. They are not metaphors. They are records of rupture orientation and thermal ancestry. Each ellipse is a fossil of expansion, a cooling scar that encodes the direction, force, and field asymmetry of its origin event.

In the recursive cosmos, ellipses proliferate not by abstraction but by necessity. They are the natural solution to spatial channel reconfiguration under scalar tension. To read an ellipse is to read a moment of cosmogenic replication.

In classical cosmology, the ellipse is often relegated to the domain of planetary motion and orbital mechanics—an elegant but ultimately descriptive tool for modeling gravitational systems. In the Unified Field Thermometry (UFT) framework, however, the ellipse undergoes a radical ontological reassignment: it is not merely a geometrical artifact, but a thermodynamic scar—visible residues of recursive rupture events that define cosmogenic replication.

Ellipses emerge wherever channel curvature fails to maintain equilibrium symmetry under scalar thermal tension. Each rupture, initiated by temperature asymmetry within the structured ether, releases not a random distortion but a directed, anisotropic fold in field geometry. This fold propagates as a thermodynamic impulse, leaving behind a spatial curvature imprint that we perceive as elliptical. Thus, ellipses are not drawn—they are inscribed, burned into the vacuum as a structural record of instability.

This elliptical signature encodes more than geometry; it records the energy gradient, directionality of rupture, and local Ξ-value at the moment of thermal destabilization. Each elliptical formation is therefore a signature not only of past rupture but of the scalar dynamics that governed it. These scars are hierarchical: one ellipse may embed others within it, as nested systems of thermal deformation interact recursively, much like geological strata encoding the history of planetary trauma.

From galaxy arms to molecular orbital patterns, the elliptical form persists across scales. Its recurrence is not incidental, but diagnostic—it confirms the recursive cosmogenesis model proposed in UFT. Just as a tree bears rings of age and trauma, the cosmos bears elliptical scars of its thermometric unfolding. Where general relativity sees curvature as consequence of mass-energy density, UFT sees it as the fossilized trace of scalar rupture—evidence of energetic differential that has since metastabilized into structure.

This view allows the ellipse to function diagnostically. By decoding the thermal signature embedded within elliptical geometries, it may be possible to infer not only prior rupture events but also the scalar field conditions (Ξ-values) that preceded them. This opens the possibility for a cosmological forensics: a reconstruction of the universe's replication history through the analysis of embedded curvature.

The ellipse is, then, neither symbol nor simplification. It is an ancient scar—etched into the fabric of space by the thermometric hammer of cosmogenic recursion. It marks not orbits, but origins. Not motion, but memory. It testifies that the universe does not expand from a single event, but replicates through nested ruptures, each encoding its own elliptical truth.

Summary

The Recursive Cosmos, as modeled by UFT-GEL, redefines the architecture of the universe as a scalar fractal—a geometry built not from initial explosion but from nested thermal ruptures. Time flickers where rupture occurs. Ellipses echo the shape of that rupture. Singularities multiply not from chaos, but from scalar recursion. ξ evolves as the local metric of becoming.

This cosmos is alive with memory, not momentum. Structured not by force, but by asymmetry. It does not flow. It replicates.

Stephen Hawking, in *A Brief History of Time*, proposed a singular, temporally-bounded origin to the cosmos: a point of infinite density, governed by relativistic constraints and quantum unpredictability. This singularity, while mathematically elegant, enforces a temporally linear model of emergence. In contrast, the UFT-GEL framework dismantles the necessity of a single, origin-point Big Bang. Instead, it posits an infinite fractal recurrence of thermodynamic initiation events, each seeded within spatial geometry by thermal disequilibrium and geometric rupture.

In the UFT-GEL formulation, time does not emerge from a singular event, but from localized scalar instabilities within a structured field. Each rupture represents not a branching from a singularity but a recursive thermodynamic pulse igniting space into new coherence. These pulses are not uniformly distributed—they are emergent, reactive to scalar density, curvature gradients, and thermal differentials across space. They replicate, not expand.

This means that every thermodynamic initiation is a localized cosmos, seeded by a Ξ-state imbalance, in which energy, temperature, volume, and curvature interact non-linearly. The recursive nature of this dynamic mimics a self-replicating geometry—a field that folds, fractures, and propagates new initiation centers via thermal echo, not explosive expansion.

Here, entropy is not a general diffusion outward but a channelized progression inward and outward through spatial geometry. The consequence is a cosmos that is not one universe, but a nested manifold of ruptural events—each igniting its own thermal topology, regulated by the local scalar Ξ.

In this view, the microwave background radiation is not a universal relic of a primordial blast, but the aggregated entropic trace of innumerable nested initiations. Inflation is thus no longer a necessary patch for early homogeneity; homogeneity is a function of recursive spatial nesting, thermally modulated, geometrically constrained.

The "start" of our observable cosmos is simply one thermal pulse among many, and its expansion is not a singular directional event but an echo of recursive ignition. There is no original flame—only the continual lighting of fires across an infinite scalar field.

This chapter redefines cosmic origin not as a boundary, but as a recurrent thermodynamic function of scalar instability. Time begins where temperature ruptures geometry. The universe is not a once-lit path, but a candle relit in every flicker of Ξ.

CHAPTER EIGHTEEN

XVIII

ιη IOTA-ETA

י״ח YOD-CHET

• • •

WAX-AK-LAJUN

DIECIOCHO

18

– A WORLD PER PULSE

18.1 The Collapse of Linear Temporality

In standard models, time is presumed to flow—smoothly, unbroken, as an axis traced across an infinite background. This assumption, inherited from classical intuition and mathematized in Newtonian and relativistic frameworks, has led to profound but ultimately flawed extrapolations about cosmic evolution, entropy, and causality. But under the UFT-GEL schema, time is not a line. It is an ignition phenomenon. It begins, pauses, ignites again—quantized not in particle form but in elliptical thermodynamic pulses.

These pulses are neither purely geometric nor simply energetic—they are thermometric curvature events. The recurrence of such events does not suggest a fixed rhythm but a recursive frequency spectrum regulated by internal field asymmetries. The quantization is elliptical because the process is not symmetric. No pulse returns to its starting condition; each is a skewed resurrection of the last. Time does not loop, but it returns obliquely.

I. Defining the Elliptical Quantization Framework

The formal expression begins with the behavior of Ξ, the universal spacetime constant:

Ξ=ET·V·πr3Ξ = \frac{E}{T \cdot V \cdot πr^3}Ξ=T·V·πr3E

In this formulation, Ξ acts not merely as a scalar value but as a *field recurrence modulator*. It defines the structural tension across a bounded spatial node at the moment of rupture. When temperature crosses a critical gradient, the system destabilizes, forming a rupture pulse. This pulse is elliptical in two critical senses:

Geometric Asymmetry — The energy influx and outflux do not trace a perfect sine or circular expansion-contraction cycle. The expansion axis and compression axis diverge, leaving a residual eccentricity encoded in spatial topology.

Temporal Skew — Each rupture alters the local temporal frequency. Time is not a constant per system but is reset—reignited—with each rupture, establishing a discrete **Ξ-state** that governs how fast or slow temporal experience proceeds within that domain.

Thus, elliptical quantization refers to the **discrete packetization of time** based on nonuniform rupture pulses—each pulse bearing its own embedded geometry and thermodynamic fingerprint.

I. Temporal Domains as Spatial Ellipses

Traditional spacetime intervals define time as a path traversed by matter. But under this model, **time is a domain**—a bounded thermodynamic geometry formed by recursive rupture. These domains are ellipsoidal because rupture is not isotropic. Field resistance, spatial channel tension, and temperature anisotropy ensure that no rupture expands uniformly.

Each temporal domain is therefore seeded as a **nested ellipse**—not metaphorically but literally: space is shaped elliptically in its infancy, encoding thermal memory in its curvature. These ellipses overlap, creating semi-stable pockets in which time behaves statistically. This also allows multiple "worlds per pulse"—a multiform expansion across fractured but contiguous domains, each obeying its own temporal logic.

II.A Quantized World Clock: Not Uniform, but Recursive

The universe does not tick forward in seconds—it pulses in elliptically-skewed quantizations. These pulses are synchronized not by an external clock but by rupture tension across channel lattices. A few consequences emerge:

Nested Temporalities: Events do not occur on one global time axis. Instead, every spatial domain has its own rupture frequency—its own clock—set by the local Ξ-state.

Pulse-Induced Multiversality: Every pulse defines a distinct thermal-geometric initiation. In UFT-GEL, this is not the branching of timelines as in quantum speculation, but the literal **creation of a new thermodynamic domain.** A new world.

Temporal Reframing of Entropy: Entropy does not merely increase—it is expressed through the shape of each ellipse. The more asymmetric the rupture, the more skewed the ellipse, the more discontinuous the experienced time within that domain.

III,Mathematical Proposal: Elliptical Pulse Function (EPF)

Define each time pulse as:

τn=f(Ξn,en)\tau_n = f(Ξ_n, e_n)τn=f(Ξ n,en)

τn=f(Ξn,en)

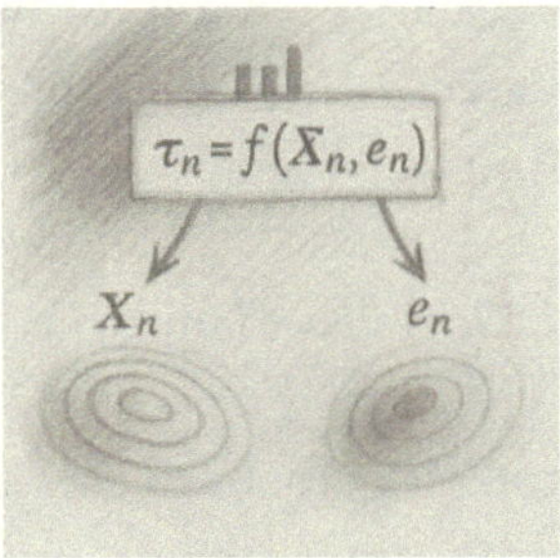

Where:

- τn\tau_nτn is the duration of the nth pulse
- ΞnΞ_nΞn is the universal field constant within that pulse
- ene_nen is the ellipticity (eccentricity) of the rupture event
- Then, time across a sequence of ruptures is not linear but a sum of discrete τ-values:

total=∑n=oNτnt_{\text{total}} = \sum_{n=o}^{N} \tau_nttotal =n=o∑Nτn

ttotal=n=0∑Nτn

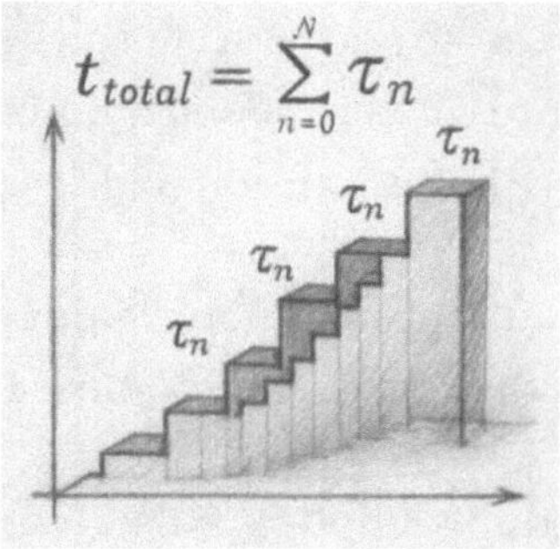

Each τ is a statistical, thermodynamic recurrence—not an inertial duration. This function can be simulated numerically if the Ξ-field curvature and boundary tensions are known.

IV. Implications for Cosmology and Quantum Systems

The Age of the Universe Is Domain-Relative: If each domain has a unique rupture recurrence and ellipticity, then "age" is a localized concept. There is no universal time, only layered thermal scars.

Quantum Flicker = Pulse Boundary Noise: At microscopic scales, the apparent unpredictability of particle position may result from the boundary tension between elliptical pulses, giving rise to "temporal flicker" in observations.

Black Hole Information = Pulse Collapse: Rather than being erased, information is embedded in the collapsing ellipse of a final pulse—flattened, not lost.

V. Toward the End of Time: Ellipticity Approaches Zero

If temperature gradients flatten and rupture ceases, then elliptical quantization collapses to a perfect sphere—a final state of thermodynamic silence. At this boundary, time no longer flickers. It vanishes. The ellipse has no tension to drive pulse recurrence.

This gives rise to a radical conclusion: **Time is not stretched—it is curved by rupture. When rupture ends, time ends.**

18.2 — Time Proceeds by Elliptical Pulse, Not Linear Flow

Time, as redefined by the UFT-GEL framework, is not a smooth, uninterrupted river. It is a sequence of discrete pulses, elliptical in geometry, propagating not uniformly, but recursively. The conventional view of time as a linear axis is a useful abstraction, but it dissolves under closer thermodynamic scrutiny. In this revised ontology, time becomes a product of rupture events that ignite structured reconfigurations within a bounded thermometric lattice.

Each pulse of time is born from the collapse and re-expansion of spatial channel geometries, seeded by localized asymmetries in temperature and curvature. These pulses are not spherical or random; they are elliptical due to the anisotropic nature of rupture—an inherent asymmetry in energy dispersal. The elliptical form arises because no rupture is perfectly centered; every ignition has bias, every implosion has a favored axis, and every expansion reflects a directional entropy trajectory.

Elliptical pulses mean that spacetime is inherently lopsided, always favoring forward causal orientation while embedding memory within the structure of each pulse. This pulse geometry sets the rhythm of emergence: not a ticking clock, but a flickering field.

Thermal disequilibrium drives each pulse into existence. The greater the temperature gradient, the tighter the ellipse, and the shorter the recurrence interval. Conversely, near-flat thermometric zones yield elongated, almost static ellipses. This correlation suggests that the rate of time is geometrically tied to the eccentricity of each pulse.

Time, therefore, is not a backdrop but an active propagation of ruptures along structured channels. Each elliptical cycle demarcates a localized event horizon, defining a bounded world-state. Between pulses, nothing moves—there is no change, no entropy, no force transmission. Only during the pulse does the universe refresh, reconfigure, and resume motion. This gives rise to a flickering, granular universe built from thermodynamic ignition events.

Elliptical quantization of time does not violate continuity; it redefines it. Continuity is an illusion sustained by high-frequency rupture cycles. At sufficiently low energies or in flattened geometries, the flicker becomes perceivable. Time slows, and the quantum of duration reveals its elliptical shape.

This understanding has implications for extreme conditions: near absolute zero, near singularities, and across cosmogenic voids.

Time halts not because it ceases to exist, but because its pulses become so distant, so diffused, that the recursive architecture suspends activation. In such regions, spacetime does not age—it waits.

To speak of time, then, is to speak of a patterned lattice of elliptical propagation events seeded by rupture, regulated by temperature, and directed by structured spatial curvature. Each pulse is a breath of becoming, and the universe is its exhale.

18.3 Each Pulse Is a Discrete World Event

Time, in the context of Project GEL and the UFT-GEL scalar field framework, is not a linear corridor but a quantized ignition — a sequence of discrete ignition events. Each temporal unit, each "pulse," is not merely a measure but a manifestation: a rupture-point that seeds its own local reality. Time is not stretched; it is locally generated, seeded by rupture, and governed by the scalar equation $\Xi = E / (T \times V \times \pi r^3)$. Each pulse configures an enclosed, self-defining world event with its own internal spatial, thermodynamic, and electromagnetic parameters.

To call each pulse a "world event" is not metaphor. In scalar thermodynamics, rupture produces curvature, and curvature forms a bounded channel. This bounded channel — thermalized and differentiated — behaves as a temporal container. Within that bounded space, laws crystallize, particles assume quantized states, and directional entropy gains traction. The world defined by that pulse is topologically enclosed yet informationally interlinked through adjacent ruptures.

Just as quantum mechanics constrains observation within probabilistic domains, so too do elliptical pulses constrain the memory and directionality of events. Events are not smeared across a timeline; they are pinned into discrete frames, with Ξ acting as the governing index of transformation. Pulse-to-pulse transitions constitute the only true "motion." The illusion of continuous flow arises from the recursive sequencing of rupture events — each encoding its own thermal, spatial, and informational content.

Importantly, these world events are not merely temporal slices of a four-dimensional manifold. They are phase-shifted instantiations, each forming a whole — a complete microcosm with defined entropy, light curvature, and vacuum modulation. Their borders are not spatial but thermodynamic. That is, a pulse begins where disequilibrium begins and ends where entropy gradients flatten. Thus, when entropy is reset — via cooling, collapse, or curvature nullification — the world defined by that pulse terminates. Another must be ignited anew.

The recursive cosmos operates not on a grand singular flow, but on a flickering logic: one pulse, one world. Within each: energy is measured, entropy increases, time flows, and light curves — until Ξ reconfigures. And then: rupture again. A new world ignites.

Every pulse, then, is not merely a duration — it is a world entire.

Each pulse is a discrete world event. It is not merely a timestamp or a position on a continuous timeline—it is a thermodynamic occurrence, an existential node, a rupture in the vacuum architecture that gives rise to a full spatial-temporal instance. Every such event unfolds from a rupture-induced configuration:

geometry, energy, entropy, and vacuum curvature interact momentarily and locally to instantiate a reality-bubble—a bounded, coherent spacetime subdomain.

Time in this view is not a river but a flickering—each flicker a cosmos, each event a pulsed ontology. This transforms conventional temporality from a unidirectional arrow to a lattice of elliptical initiations, where recurrence and instability govern succession.

Ξ defines the inner logic of every seeded bubble. It is the scalar fingerprint that encodes energy density, thermal differential, volumetric tension, and spatial curvature within that pulse. Ξ is not merely a parameter—it is the pulse's governing equation. It determines whether a rupture will yield coherence or collapse, resonance or decay. Like a DNA strand in biology, Ξ embeds the thermodynamic instructions for the pulse's structural evolution.

In any given seeded bubble, Ξ governs the symmetry or asymmetry of curvature, the rate of entropy propagation, and the effective gravitational and electromagnetic coupling constants. Variations in Ξ across bubbles produce distinct cosmological signatures, permitting the possibility of multivariate universes—each with unique spacetime fabrics shaped by its initiating rupture.

Thus, seeded bubbles are not scaled replicas but Ξ-modulated variations—each a discrete cosmos shaped from within. Where Ξ is stable and non-zero, continuity ensues; where Ξ approaches divergence, rupture follows, seeding the next generation. This recursive architecture enables a theory not just of time, but of universe reproduction via thermodynamic ignition and scalar regulation.

This defines the recursive cosmos—not as an infinite spread of homogeneous space, but as a self-regulating thermodynamic matrix of pulsed, Ξ-shaped realities.

18.4: Ξ Defines the Inner Logic of Every Seeded Bubble

Within the framework of Project GEL and the thermodynamic field geometry it describes, each cosmogenic event—each pulse of spacetime—is not merely a product of energy release or vacuum fluctuation. Rather, each event constitutes a bubble whose internal behavior is governed by a scalar thermodynamic logic. That logic is encoded in the Universal Spacetime Constant: $\Xi = E / (T \times V \times \pi r^3)$. Far from being an abstract ratio, Ξ emerges as the core blueprint regulating the properties of discrete world domains.

Ξ governs the internal symmetry, temporal granularity, and thermodynamic inertia of each seeded bubble. It determines not only how fast entropy can evolve, but whether it evolves at all. Lower Ξ-values imply sluggish thermodynamic curvature, flat gradient conditions, and frozen time—a near-static interior in which quantum potential cannot unfold. High Ξ-values, by contrast, signal steep thermal gradients and spatial tension capable of sustaining entropy-driven dynamism. The bubble becomes temporally vivid, causally rich, and evolutionarily potent.

This internal logic is what differentiates one bubble from another, even if they emerge from similarly scaled ruptures. Two regions with equal energy inputs but different confinement volumes or radii will express entirely distinct Ξ profiles—and thus different temporal, structural, and informational outcomes. The seeded bubble is not randomly formed; it is thermodynamically coded. Its potential for light propagation, matter coalescence, and entropy evolution is written into its Ξ-state at birth.

From a geometric standpoint, this means that space is not uniformly elastic. Expansion or collapse is not driven by kinetic force alone but by the internal pressure logic dictated by Ξ. The scalar integrates thermodynamic, volumetric, and spatial curvature data into one unifying coefficient. In this sense, Ξ functions as both the DNA and the governor of any localized spacetime construct.

Moreover, within each bubble, the reconfiguration of internal frames—temporal pulses, matter clustering, light vectors—unfolds recursively under the constraints imposed by the bubble's defining Ξ-value. This determines the permissible rhythms of quantized time, the allowed propagation speeds for information, and the entropic directionality of internal processes. The inner logic is scalar-absolute, not observer-relative.

Insofar as these seeded bubbles define local cosmoses, the multiverse can no longer be understood as an ensemble of disconnected possibilities but as a spectrum of Ξ-structured thermodynamic geometries. Each bubble is a semi-closed thermodynamic argument: a proposition encoded in heat, volume, curvature, and structure. It is not the explosion that gives it identity—it is the scalar discipline within the bubble that dictates its fate.

Thus, Ξ is not merely a cosmological constant or a scaling convenience. It is the logic by which thermal rupture becomes structured, motion becomes permissible, and time becomes meaningful. The seeded bubble is a Ξ-governed architecture—a thermometric world in pulse.

18.5 — Time Is Not Stretched: It Is Ignited

The pulse has come full ellipse.

What began in recursive thermodynamic rupture now concludes in ignition—a sequence of seeded asymmetries, not in continuum, but in curvature. Time, as now revealed, is neither the arrow nor the thread, but the flame that scars the vacuum. It is neither cosmic clockwork nor relativistic drift—it is detonation structured by gradient, geometry written by pressure.

Chronus is not a neutral backdrop. It is an effect, a signal, a symptom of rupture. Every pulse, every discrete elliptical ignition, defines a world—not metaphorically, but causally. Ξ, the Universal Spacetime Constant, anchors each temporal emergence to a quantifiable scalar logic: energy divided by thermometric spatial asymmetry. Time, therefore, is not a vector across which the cosmos moves. It is the ignition event through which the cosmos exists at all.

Across these chapters, we have dismantled the illusion of time as a smooth arrow. It does not stretch. It does not flow. It flickers, ruptures, ignites. Implosions generate voids, and these voids function as scalar pistons—generating structure through the collapse of equilibrium. Each moment is not a passing; it is a thermodynamic re-creation. Each frame of existence is a seeded bubble, defined by local curvature, regulated by temperature, bound within elliptical recurrence.

Time is born from failure—of symmetry, of stillness, of total equilibrium. In that collapse arises a metric, a geometry, a domain. The world is not continuous; it is discretely pulsed. A world per pulse, a cosmos per scar. Each ellipse is not the echo of origin but the origin itself—iterative, nested, recursive.

The ancient myths intuited what our metrics now confirm. Time, like Chronos, is not gentle. It divides. It reaps. And through division, it seeds.

We have entered not a new story of time, but a new structure. Chronos is now defined.

Not as duration—but as rupture. Not as measure—but as ignition. Not as passage—but as birth.

Ξ governs the spark. The vacuum yields to asymmetry. And time, scar by scar, becomes space made visible.

This is not the end. This is another ignition. (. ..)

'FIN

Time as Elliptic Seeding and Temporal Replication

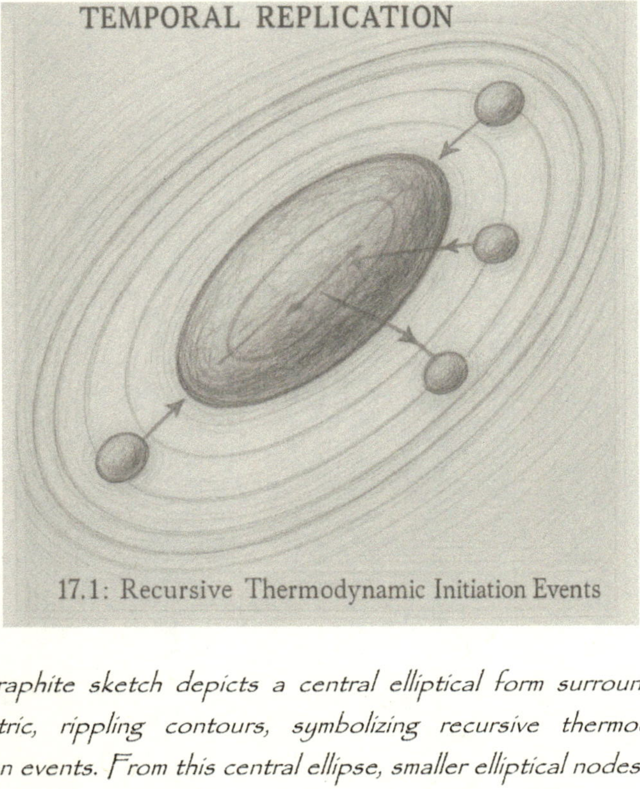

17.1: Recursive Thermodynamic Initiation Events

The graphite sketch depicts a central elliptical form surrounded by concentric, rippling contours, symbolizing recursive thermodynamic initiation events. From this central ellipse, smaller elliptical nodes radiate outward along the ripples, each connected by directional arrows, representing nested ruptures and temporal replication. The composition conveys dynamic motion and layered geometry, embodying the concept of time as flickering, elliptical seeding.

Epilogue

— The Rupture Behind All Things

In the end, time does not unroll—it erupts. It is not the loom of continuity but the scar of ignition, not the backdrop of being but its first rupture. What we call "past" is not a receding trail, but a stratification of thermal events, frozen in ellipses. Each moment is not a unit of linear progression, but a discrete flame within the vacuum: structured, pulsed, scarred.

This work has traced a path not forward but inward. Not along a timeline, but through the layered caverns of thermometric memory. If Hawking described the curvature of spacetime, this volume has unveiled the thermometric engine behind that curve. Ξ was not introduced as symbol, but as syntax: the scalar grammar of rupture, the invariant residue left where equilibrium fails and a world ignites.

We have shown that light is not a vector of velocity but of entropy. That expansion is not detonation but piston logic. That time does not stretch—it flickers. That behind every world is a wound, and behind every wound, a structure. We did not travel time—we seeded it. We did not observe flow—we measured ignition. This is the cosmos as recursive scar, as nested ellipse, as the memory of asymmetry encoded in vacuum geometry.

Time is the breath of rupture. The cosmos does not grow older. It pulses. Let this be the final word of this volume, and the first breath of what must follow:

Time began when stillness broke. And it breaks still.

Bibliography

Primary Works in Physics and Cosmology

- Einstein, Albert. *Relativity: The Special and General Theory*. New York: Crown Publishers, 1961.
- Hawking, Stephen. *A Brief History of Time*. New York: Bantam Books, 1988.
- Planck, Max. *The Theory of Heat Radiation*. New York: Dover Publications, 1959.
- Wheeler, John A., and Kenneth W. Ford. *Geons, Black Holes, and Quantum Foam: A Life in Physics*. New York: W. W. Norton, 1998.
- Feynman, Richard P., Robert B. Leighton, and Matthew Sands. *The Feynman Lectures on Physics, Vol. 1–3*. Reading, MA: Addison-Wesley, 1963.
- Callen, Herbert B. *Thermodynamics and an Introduction to Thermostatistics*. 2nd ed. New York: Wiley, 1985.
- Weinberg, Steven. *The Quantum Theory of Fields*. Cambridge: Cambridge University Press, 1995.

Alternative Cosmologies, Catastrophism, and Planetary History

- Velikovsky, Immanuel. *Worlds in Collision*. New York: Macmillan, 1950.
- Velikovsky, Immanuel. *Earth in Upheaval*. New York: Doubleday, 1955.
- Velikovsky, Immanuel. *Ages in Chaos*. New York: Doubleday, 1952.
- Velikovsky, Immanuel. *Peoples of the Sea*. Garden City, NY: Doubleday, 1977.
- Velikovsky, Immanuel. *Ramses II and His Time*. Garden City, NY: Doubleday, 1978.

- Velikovsky, Immanuel. *Oedipus and Akhnaton.* Garden City, NY: Doubleday, 1960.
- Velikovsky, Immanuel. *Mankind in Amnesia.* Garden City, NY: Doubleday, 1982.
- Larsen, Anthony. *The Moon Shall Turn to Blood: Prophecy, Catastrophism and the Signs in the Heavens.* Salt Lake City: Millennial Star, 1980.
- Larsen, Anthony. *And the Sun Shall Be Darkened: The Signs in Heaven Before the Great and Terrible Day of the Lord.* Salt Lake City: Millennial Star, 1983.
- Larsen, Anthony. *There Shall Be a New Heaven and a New Earth.* Salt Lake City: Millennial Star, 1985.
- Talbott, David. *The Saturn Myth.* New York: Doubleday, 1980.
- Talbott, David, and Wallace Thornhill. *The Electric Universe.* Portland, OR: Mikamar Publishing, 2007.

Ancient Astronomy, Mythology, and the Venus Golden Age

- Cardona, Dwardu. *Flare Star: The Mythology of the Venus Golden Age.* Victoria, B.C.: Trafford Publishing, 2006.
- Aveni, Anthony F. *Skywatchers of Ancient Mexico.* Austin: University of Texas Press, 1980.
- Krupp, E. C. *Echoes of the Ancient Skies: The Astronomy of Lost Civilizations.* New York: Harper & Row, 1983.
- Kelley, David H., and Eugene F. Milone. *Exploring Ancient Skies: A Survey of Ancient and Cultural Astronomy.* 2nd ed. New York: Springer, 2011.

Scripture and Theological Works

- The Holy Bible, King James Version.

- The Book of Mormon: Another Testament of Jesus Christ.
- The Doctrine and Covenants of the Church of Jesus Christ of Latter-day Saints.
- The Book of Abraham, The Pearl of Great Price of the Church of Jesus Christ

Philosophy of Time and Thermodynamics

- Barbour, Julian. The End of Time: The Next Revolution in Physics. Oxford: Oxford University Press, 1999.
- Rovelli, Carlo. The Order of Time. New York: Riverhead Books, 2018.
- Prigogine, Ilya. From Being to Becoming: Time and Complexity in the Physical Sciences. San Francisco: W. H. Freeman, 1980.

About the Author

Miguel A. Tinoco is an independent researcher, visionary thinker, theologian, and theoretical physicist whose work boldly challenges convention and bridges the worlds of science, theology, and cosmopolitical thought. Known for his unapologetic and original insights, Tinoco invites readers to explore the deepest questions of existence—where thermodynamics, time, ancient myth, and scripture converge in a unified vision of reality.

A self-educated scholar, Tinoco left formal schooling after the 10th grade, earning his GED from High Point High School at an early age. Choosing a path of practical engineering, family devotion, and priesthood over conventional academic accolades, he committed himself to lifelong, rigorous independent study in astrophysics, cosmology, theology, and advanced mathematics. That discipline continues to fuel his work today.

Ordained as a Deacon, Teacher, and Priest in the Aaronic Priesthood beginning at age 12, and later as an Elder in the Melchizedek Priesthood, Tinoco has served with unwavering faith, maintaining his credentials and standing before God. His writings are deeply shaped by this covenantal foundation and his work goes before him for he has been writing his theological and theoretical insights for decades.

Tinoco is the architect of the Unified Field Theory of Gravitational-Electromagnetic-Light (UFT-GEL) and Project GEL, proposing the Universal Spacetime Constant (Ξ)—a bold integration of physics and scripture, rooted not in idle speculation, but in disciplined study, revelation, and the pursuit of truth. Each work and volume is an ever expanding construct of his original mission and work.

He is the author of four major works:

- *10 Ways to Stand Like Michael*
- *Healing in Gilead*
- *Scientia Unita Universum*
- *Fractura Temporis*

In addition to these books, Tinoco has published over 30 blogs containing more than 1,000 posts on subjects ranging from theology and science to cosmopolitan commentary. His voice continues to inspire those who seek harmony between faith and reason, tradition and innovation. Like the approach valued in Japanese and Chinese learning, he holds that science, mathematics, or even theology, if not developed and explained with plainness, rooted in real work experience, and grounded in self-evident truth and tested fact, remains a science still in its infancy.

Acknowledgment of the Role of AI

The author gratefully and humbly acknowledges the use of artificial intelligence and Grammarly as editing tools in the preparation of this work—not as a source of original thought or regenerative conceptual development, but as a laboratory for testing and contradicting ideas, a tool for extensive editing and proofreading, and an aid in formatting, structuring, and refining the manuscript and captioning. The insights, theories, and formulations presented herein are the product of independent study, reflection, and revelation. The role of AI was purely instrumental: to assist in organizing the material, improving clarity, and ensuring technical precision where appropriate. All conclusions, interpretations, and innovations belong solely to the author.

www.ingramcontent.com/pod-product-compliance
Lightning Source LLC
LaVergne TN
LVHW100726061126
851857LV00011B/410